Image Processing with Python

A practical approach

Online at: https://doi.org/10.1088/978-0-7503-5924-5

Image Processing with Python

A practical approach

Edited by
Irshad Ahmad Ansari
ABV-Indian Institute of Information Technology and Management Gwalior, Gwalior, India

Varun Bajaj
Maulana Azad National Institute of Technology, Bhopal, India

IOP Publishing, Bristol, UK

© IOP Publishing Ltd 2024

ISBN   978-0-7503-5924-5 (ebook)
ISBN   978-0-7503-5922-1 (print)
ISBN   978-0-7503-5925-2 (myPrint)
ISBN   978-0-7503-5923-8 (mobi)

DOI   10.1088/978-0-7503-5924-5

Version: 20240701

IOP ebooks

British Library Cataloguing-in-Publication Data: A catalogue record for this book is available from the British Library.

Published by IOP Publishing, wholly owned by The Institute of Physics, London

IOP Publishing, No.2 The Distillery, Glassfields, Avon Street, Bristol, BS2 0GR, UK

US Office: IOP Publishing, Inc., 190 North Independence Mall West, Suite 601, Philadelphia, PA 19106, USA

Contents

Anil B Gavade, Rajendra B Nerli, Pushkar Bansidhar Patil, Richa Ravi Siddannavar, Venkata Siva Prasad Bhagavatula and Priyanka A Gavade

Nikhil Kushwaha, Om Asati and Mainak Sadhya

Preface

Python for image processing applications has grown tremendously due to its open-source nature and excellent library support. Professionals, students, and researchers working in the image processing domain find it helpful to use pre-existing libraries. They can even create their own library and make it available for others to use. These qualities led to an excellent growth in the popularity of Python.

Image processing is quite a complex computation. This can be done efficiently by using matrix-based calculations. In addition, development in machine learning and artificial intelligence domains also helped the image processing domain grow significantly. In the past, Python has been proven to be an excellent platform for this combination, i.e., machine learning-supported image processing.

Python provides many pre-trained machine-learning models. These models can easily be used for different image processing tasks with a small amount of transfer learning. Applications such as image de-noising, character recognition, and image sharpening are a few domains that get affected by this upgrade.

The proposed work attempts to bring together recent trends in the image processing domain and provide hands-on exposure to the readers, students, and researchers working in this domain. This book contains an example-led explanation of image processing applications and machine learning models. Python is used for the implementation work to make this book useful for the masses.

Chapter 1 introduces the different Python libraries and their applications for the image processing domain. Chapter 2 provides a methodological study of image processing operations such as de-noising, image segmentation, edge detection, object recognition, image sharpening, classification, feature detection and matching, analysis, and manipulation of images. Chapter 3 presented a step-by-step process for implementing three different image segmentation techniques using Python. Chapter 4 explored the processes to improve the segmentation of different regions of the images. Therefore, a pre-processing such as adaptation was performed on test images, and then a region-growing algorithm was performed using 8-neighborhood to generate new pixels. Chapter 5 discusses the principles of retinal layer segmentation and elaborates on the design cycle of the deep learning methods for retinal layer segmentation in OCT images. Chapter 6 provides de-noising based on the wavelet soft thresholding technique. Chapter 7 provides comparative evaluations of various deep learning architectures, including U-Net, 3D U-Net, V-Net, and nU-Net, revealed that nnU-Net exhibited superior performance in accurately segmenting prostate cancer within the peripheral zone (PZ) and central gland (CG) regions. Chapter 8 describes the working of the Optical Character Recognition to extract the text content of the Image and the implementation of the Convolutional Neural Network. Chapter 9 classifies the coronavirus disease (COVID-19) with computed tomography (CT) images based on artificial intelligence (AI) and a binary neural network (BNN). Chapter 10 reviews significant image despecking methods developed in the past. Chapter 11 addresses the various data collection, pre-processing, feature extraction, model training, and prediction stages involved in precision agriculture utilizing image processing and machine learning.

Acknowledgements

Dr Ansari expresses his gratitude and sincere thanks to his wife, family members and teachers for their constant support and motivation.

Dr Bajaj expresses his heartfelt appreciation to his mother Prabha, wife Anuja, and daughter Avadhi, Avyana for their wonderful support and encouragement throughout the completion of this important book on Image Processing with Python: A Practical Approach. His deepest gratitude goes to his mother-in-law and father-in-law for their constant motivation. This book is an outcome of sincere efforts that could be given to the book only due to the great support of the family.

We sincerely thank Prof. S N Singh, Director of ABV-IIITM Gwalior and Professor Karunesh Kumar Shukla, Director of Maulana Azad National Institute of Technology (MANIT) Bhopal for their support and encouragement. We would like to thank all our friends, well-wishers and all those who keep us motivated to do more and more; better and better. We sincerely thank all contributors for writing relevant theoretical background and applications of this book.

We humbly thank Dr John and all editorial staff of IOP for their excellent support, necessary help, appreciation, and quick responses. We also wish to thank IOP for giving us this opportunity to contribute to a relevant topic with a reputable publisher. Finally, we want to thank everyone, in one way or another, who helped us edit this book.

Dr Bajaj especially thanks his family who encouraged him throughout the time of the editing of this book. This book is heartily dedicated to his father who took the lead to heaven before the completion of this book.

Last but not the least, we would also like to thank God for showering us with his blessings and strength to do this type of novel and quality work.

Irshad Ahmad Ansari
Varun Bajaj

Editor biographies

Irshad Ahmad Ansari

Irshad Ahmad Ansari (PhD, SMIEEE20) has been working as an Assistant Professor Grade I in the Department of Electrical and Electronics Engineering at ABVIIITM, Gwalior, India, since June 2023. He received his Bachelor of Technology degree in Electronics and Communication Engineering from Gautam Buddh Technical University (formally Uttar Pradesh Technical University), Lucknow, India, in 2010 and his Master's of Technology (M.Tech.) degree in Control and Instrumentation from Dr B R Ambedkar National Institute of Technology (NIT) Jalandhar, Punjab, India in 2012. He completed his PhD at the Indian Institute of Technology Roorkee with Ministry of Human Resource Development (MHRD) teaching assistantship in 2017 and subsequently joined the Gwangju Institute of Science and Technology, South Korea as a postdoctoral fellow. Afterward, he joined PDPM IIITDMJ as Assistant Professor Grade II. His major research interests include signal and image processing, electronic design, ML, biomedical signal processing, computer vision etc. He is contributing as an active technical reviewer of leading international publishers such as IEEE, IOP Publishing, Elsevier, and Springer. He has more than 70 publications, which include 29 SCI/SCIE journal papers, 28 international conference papers, 4 edited books, and 6 book chapters. The citation impact of his publications is around 1200 citations, with an h-index of 19, and an i10 index of 27 (Google Scholar December 2023). He has guided three (3 awarded) PhD scholars and 14 M. Tech scholars. He has been listed as the world's top 2% of researchers/scientists by Stanford University, USA (October, 2023).

Varun Bajaj

Varun Bajaj (PhD, SMIEEE20) is working as an Associate Professor in the discipline of Electronics and Communication Engineering, at Maulana Azad National Institute of Technology Bhopal, India from Jan 2024. He served as Associate Professor in the discipline of Electronics and Communication Engineering, Indian Institute of Information Technology, Design and Manufacturing (IIITDM) Jabalpur, India since July 2021 to Jan 2024. He worked as an Assistant Professor at IIITDM Jabalpur from March 2014 to July 2021. He also worked as a visiting faculty in IIITDM Jabalpur from September 2013 to March 2014. He worked as Assistant Professor at the Department of Electronics and Instrumentation, Shri Vaishnav Institute of Technology and Science, Indore, India during 2009–2010. He received his PhD degree in the Discipline of Electrical Engineering, at the Indian Institute of Technology Indore, India in 2014. He received an MTech degree with Honors in Microelectronics and VLSI design from Shri Govindram Seksaria Institute of Technology and Science, Indore, India in

2009, and a BE degree in Electronics and Communication Engineering from Rajiv Gandhi Technological University, Bhopal, India in 2006.

He is an Associate Editor of *IEEE Sensor Journal, Biomedical Signal Processing for Frontiers in Signal Processing*, Subject Editor-in-Chief of *IET Electronics Letters*. He served as a Subject Editor of *IET Electronics Letters* from November 2018 to June 2020 and Guest Editor in Elsevier and IET Journals. He is Senior Member IEEE June 2020, MIEEE 16–20, and contributing as active technical reviewer of leading international journals of IEEE, IET, and Elsevier, etc. He has served as member of the review boards for around 80 scientific journals. He has also served on the scientific committees of various international conferences. He has delivered more than 50 expert talks and lectures in conferences, workshops, and short-term courses organized by various institutes. He has received several awards including Achievement Award Best Paper Award (ICHIT conference 2012 and 2nd International Conference on Computational Electronics for Wireless Communication (Springer) 2022), and 'Award for Excellence in Research' by 7th South Asian Education Awards-Summit 22. He has been listed in the world's top 2% researchers/scientists by Stanford University, USA (October 2020, October 2021, and October 2022).

He has 170 publications, which include 104 journal papers, 35 conference papers, 13 books, and 18 book chapters. He has also been granted 2 international patents and published 3 Indian patents. The citation impact of his publications is around 5758 citations, with an h-index of 42, and i10 index of 108 (Google Scholar March 2023). He has guided 9 (6 completed, and 3 In process) PhD Scholars, and 8 MTech scholars. He has worked on 5 research projects funded by DST and CSIR and 3 Consultancy projects. His research interests include biomedical signal processing, AI in healthcare, brain computer interfaces, pattern recognition, and ECG signal processing.

List of contributors

Om Asati
PDPM Indian Institute of Information Technology for Design and Manufacturing, Jabalpur, India

Varun Bajaj
Maulana Azad National Institute of Technology Bhopal 462003 MP India

Venkata Siva Prasad Bhagavatula
Medtronic, Hyderabad, India

Siddharth Bhalerao
Gyan Ganga Institute of Technology and Sciences, India

Devanand Bhonsle
SSTC, Bhilai, India

V R Deepthi
National Institute of Technology Calicut, Calicut, India

Akanksha Dixit
PDPM IIITDM Jabalpur, MP, India

Anil B Gavade
Department of E&C, KLS Gogte Institute of Technology, Belagavi, Karnataka, India

Priyanka Gavade
Department of Computer Science and Engineering, KLE Society's Dr M S Sheshgiri College of Engineering and Technology, Belagavi, Karnataka, India

Zahra Ghanbari
Biomedical Engineering Department, Amirkabir University of Technology, Tehran, Iran

Swati Hadke
GHRCE, Nagpur, India

Atiya Khan
School of Computer Science, Dr Vishwanath Karad MIT World Peace University, Pune, MH, India

Nikhil Kushwaha
Indian Institute of Information Technology for Design and Manufacturing, Jabalpur, India

Rajveer S Lalawat
PDPM IIITMJ, Jabalapur, India

Shankar Mali
School of Computer Science, Dr Vishwanath Karad MIT World Peace University, Pune, MH, India

Ravi Mishra
GHRCE, Nagpur, India

Anupama Mohabansi
GHRCE, Nagpur, India

Sheetal Mungale
GHRCE, Nagpur, India

Manjushree Nayak
NIST(Autonomous), Berhampur, Odisha, India

Rajendra B Nerli
Department of Urology, JN Medical College, India

Chandrashekhar H Patil
School of Computer Science, Dr Vishwanath Karad MIT World Peace University, Pune, MH, India

Pushkar Bansidhar Patil
Department of E&C, KLS Gogte Institute of Technology, Belagavi, Karnataka, India

Anu G Pillai
Kalinga University, Raipur, C.G., India

Roshni Rahangdale
SSTC, Bhilai, C.G., India

Mainak Sadhya
Indian Institute of Information Technology for Design and Manufacturing, Jabalpur, India

Sourabh Sahu
Gyan Ganga Institute of Technology and Sciences, India

Amin Sakhaei
Biomedical Engineering Department, Amirkabir University of Technology, Tehran, Iran

Rishab Sarkar
National Institute of Technology Calicut, Calicut, India

Ruhi Uzma Sheikh
ACET, Nagpur, Maharashtra, India

Richa Ravi Siddannavar
Department of E&C, KLS Gogte Institute of Technology, Belagavi, Karnataka, India

G Sreelekha
National Institute of Technology Calicut, Calicut, India

P V Sudeep
National Institute of Technology Calicut, Calicut, India

Saurabh Tewari
Gyan Ganga Institute of Technology and Sciences, India

Shruti Tiwari
SSTC, Bhilai, C.G., India

Prajakta Upadhye
GHRCE, Nagpur, India

Amol D Vibhute
Symbiosis Institute of Computer Studies and Research (SICSR), Symbiosis International (Deemed University), Pune, MH, India

Contributor biographies

Om Asati

I am a student, currently in my 2nd year at PDPM Indian Institute of Information Technology, Design, and Manufacturing Jabalpur (IIITDMJ). I am pursuing a Bachelor of Technology (BTech) degree in Electronics and Communication Engineering (ECE).

I was born on September 17, 2003, in Chhatarpur, Madhya Pradesh, which is my hometown.

At IIITDMJ, I have been studying various subjects related to electronics, communication systems, and signal processing. The institute focuses on information technology, design, and manufacturing, providing me with a well-rounded education in these areas.

Being in the 2nd year of my BTech program, I have already had the opportunity to engage in academic pursuits, collaborate with fellow students, and work on hands-on projects. I have been able to explore my interests within the field of electronics and communication engineering.

I am excited about the opportunities ahead as I continue my academic journey. I hope to make significant contributions to the field and leverage my knowledge and skills to positively impact society.

Overall, I am proud to be a student of PDPM IIITDMJ, pursuing my passion for Electronics and Communication Engineering and looking forward to the future.

Venkata Siva Prasad Bhagavatula

Venkata Siva Prasad Bhagavatula is working as Principal Systems Engineer with Medtronic Innovation and Engineering Center, Hyderabad, India. Venkata Siva Prasad Bhagavatula completed his BE degree in Instrumentation Engineering from Karnataka University, Dharwad and completed Master of Science (Online) in Data science from Liverpool John Moores University, UK. He holds a patent, worked in four new product launches in various roles in Research and Development in Healthcare industry. He has 19 years of experience in the areas of hardware, systems engineering. His main area of work is in systems engineering. His main interests include medical devices, data science and machine learning applications to medical imaging.

Siddharth Bhalerao

Dr Siddharth Bhalerao is presently working as an Assistant Professor in Department of Computer Science and Engineering (CSE) at Gyan Ganga Institute of Technology and Sciences, Jabalpur, India. He has completed his PhD from PDPM Indian Institute of Technology and Sciences, Jabalpur. He has earned his Bachelor's degree in Electronics and Communication Engineering and Master's degree in Digital Communication from Rajiv Gandhi Proudyogiki Vishwavidyalaya, Bhopal. His research interests include signal and image processing, watermarking, Internet of Things (IoT), artificial intelligence and machine learning.

Devanand Bhonsle

Dr Devanand Bhonsle, PhD received his BE from Pt. RSSU, Raipur, ME and PhD degrees from CSVTU, Bhilai in Electronics and Telecommunication in 2004, 2008 and 2019, respectively. He is currently working as a Sr. Assistant professor in the Faculty of Engineering, Shri Shankaracharya Technical Campus, Bhilai from July 2005. He has 18 years of experience. His research interest includes signal processing and image processing. He authored/co-authored 30 publications in various peer-reviewed journals. He has written 8 book chapters and 1 book and presented 30 papers in national/international conferences.

V R Deepthi

V R Deepthi, received her BE degree in Computer Science and Engineering from S Veerasamy Chettiar College of Engineering and Technology, Tenkasi, Tamil Nadu, India in 2005 and MTech degree in Computer Science and Engineering from National Institute of Technology, Calicut, Kerala, India in 2015. She joined as a Lecturer in the Department of Computer Science and Engineering, College of Engineering Trivandrum, Kerala in 2009 and is currently working as Assistant Professor in the same department. Her research interests include application of machine learning and deep learning techniques in biomedical image analysis.

Akanksha Dixit

Akanksha Dixit received her BE degree in Electronics and Communication Engineering and MTech degree in Embedded system and VLSI Design from Rajiv Gandhi Technological University, Bhopal, India, in 2009 and 2012, respectively. She was working as an Assistant Professor with the Gyan Ganga Institute of Technology and Sciences, Jabalpur, India. She is currently a Research Scholar with the PDPM-Indian Institute of Information Technology, Design and Manufacturing, Jabalpur, India. Her research interests include biomedical signal processing, machine learning, and deep neural networks.

Anil B Gavade

Dr Anil B Gavade is an accomplished researcher and educator in the field of Electronics and Communication Engineering. He holds a Bachelor of Engineering degree in Instrumentation Engineering from Karnataka University, Dharwad, obtained in 2000. He further pursued his academic journey and completed a Master of Technology in Digital Electronics in 2005. In 2019, he successfully obtained his PhD from Visvesvaraya Technological University, Belagavi, specializing in the application of deep learning techniques for land use land cover classification from satellite imagery. Currently serving as an Associate Professor at the Department of Electronics and Communication in KLS Gogte Institute of Technology, Belagavi, Karnataka, India. He has contributed significantly to the academic community through publications in reputable journals and participation in various international and national conferences. His research interests primarily revolve around explainable artificial intelligence (XAI), human-centered AI, computer vision, and machine perception.

Priyanka A Gavade

Priyanka A Gavade received her BE and ME degrees in Computer Science and Engineering from Visveswaraya Technological University, Karnataka, India, in 2007, and 2009, respectively. She is currently pursuing She is currently pursuing research in the field of Image Processing and Deep Learning. She is currently a Assistant professor at KLE Technological University's Dr M S Sheshgiri College of Engineering and Technology. Priyanka Gavade has authored/co-authored five publications in various peer-reviewed, journals. Also, she has presented about two papers in international conferences.

Zahra Ghanbari

Zahra Ghanbari received her BSc in Electronics Engineering from Ferdowsi University of Mashhad, Iran, in 2005. She received her MSc in biomedical engineering from Sharif University of Technology, Tehran, Iran, in 2011. Zahra received her PhD in biomedical engineering from Amirkabir University of Technology, Tehran, Iran, in 2020. She was an adjunct professor at Sadjad University of Technology, Imam Reza International University and Kayyam Institute for Higher Education. Furthermore, she was teaching at the Electrical Engineering and Computer Engineering department, Alzahra University, Tehran, Iran (2016–21). Now she is an adjunct professor at Ferdowsi University, Mashhad, Iran, Shahid Beheshti University, Tehran, Iran, and Amirkabir University of Technology, Tehran, Iran. She has authored more than 14 journal and conference papers, in addition to 8 book chapters. Her research and teaching interests include biomedical signal and image processing, neuroscience, cognitive science, analogue and digital electronics.

Swati Hadke

Swati Hadke is the research Scholar of discipline Electronics and Telecommunication Engineering at G H Raisoni University, Amravati (Maharashtra) 444701, India. She received BE degree from RTMNU, Nagpur in 2011. She received MTech degree from RTMNU, Nagpur in 2015. She served as an Assistant Professor at GHRIETNagpur from September 2011 to till date. She has authored/co-authored 9 research papers in various reputed international and national publishers' journals/conferences. Her research interests include video shot boundary detection, embedded system and artificial intelligence, etc.

Atiya Khan

Atiya Khan is a PhD Scholar at the School of Computer Science, Dr Vishwanath Karad MIT World Peace University, Pune, Maharashtra, India. She is assistant professor at G H Raisoni College of Engineering, Nagpur, Maharashtra, India. She received her MCA (Computer Science) post graduate degree from RTMNU, Nagpur Maharashtra, India. She has 18+ years of academic experience teaching at the UG and PG levels research and innovation, and she contributed to more than 10 refereed journals, book chapters, and conference papers that were indexed by Scopus, SCIE, and UGC. Her areas of interest in research include GIS, hyperspectral-multispectral remote sensing, pattern recognition, and image processing.

Nikhil Kushwaha

I am **Nikhil Kushwaha**, an accomplished individual currently pursuing a BTech degree in Electronics and Communication Engineering (ECE) at PDPM Indian Institute of Information Technology, Design and Manufacturing Jabalpur.

I was born on April 29, 2002, in Kondagaon, located in Chhattisgarh, India. From a young age, I developed a deep fascination for electronic devices and their intricate workings, igniting a desire to explore the endless possibilities they offer. This innate curiosity and enthusiasm led me to pursue a BTech degree in ECE, enabling me to delve deeper into the realm of electronics and communication engineering.

As a student at PDPM Indian Institute of Information Technology, Design and Manufacturing Jabalpur, I am fortunate to benefit from a nurturing and supportive academic environment. The institute is widely recognized for its commitment to academic excellence, innovation, and design. I have access to cutting-edge

technology and state-of-the-art laboratories, allowing me to develop practical skills and gain valuable hands-on experience, preparing me for the challenges and opportunities that lie ahead.

In addition to my academic pursuits, I actively engage in extracurricular activities and initiatives within the institute. I am a dedicated member of various student organizations and clubs, collaborating with like-minded individuals to contribute to innovative projects. My strong interpersonal skills and ability to work effectively in teams have earned me the respect and admiration of my peers and professors alike.

Looking towards the future, my ultimate goal is to leverage my knowledge and skills to make a significant impact in the field of electronics and communication engineering. I aspire to contribute to the development of cutting-edge technologies and tackle real-world challenges. With my unwavering dedication, commitment to academic excellence, and thirst for knowledge, I am poised to emerge as a respected professional in the electronics industry.

Rajveer Singh Lalawat

Rajveer Singh Lalawat received the BE degree in electronics and communication engineering from the Govt. Indira Gandhi Engineering Institute, Sager, India, in 2013, and the MTech degree in ABV Indian Institute of Information Technology and Management, Gwalior in 2019. He worked as a design engineer at Sion Semiconductor Pvt. Limited Bengaluru Karnataka. He is currently a PhD Research Scholar at the Department of Electronics and Communication Engineering, PDPM Indian Institute of Information Technology, Design and Manufacturing, Jabalpur 482005, India (email:21pece05@iiitdmj.ac.in).

Shankar Mali

Prof. Shankar Mali is currently working as Professor at Department of Computer science, Dr Vishwanath Karad MIT World Peace University. He is also handling additional responsibility as a Director of IQAC at MIT-WPU. He has experience of 23 years teaching MCA and MSc (Computer Science) students. He has received PhD degree from Solapur University in Computer Science (2011) and qualified SET in Computer Science in year 2004. Two students have successfully completed PhD under his guidance from Savitribai Phule Pune University, Pune. He is presently guiding two PhD students at MIT World Peace University, Pune. His research areas include digital image processing, pattern recognition and natural language processing.

Ravi Mishra

Dr Ravi Mishra is the faculty of discipline of Electronics and Telecommunication Engineering at G H Raisoni University, Amravati (Maharashtra) 444701, India since July 2019. He served as Sr. Assistant Professor at SSTC from Feb 2010 to Jul 2019, Sr Lecturer at DIMAT Raipur from August 2009 to February 2010. He also served as Head-R&D at Indus Global Software Solutions from January 2006 to March 2008. He received BE degree in Electronics and Telecommunication from BIT Raipur in 2002, and the ME degree in Electronics and Telecommunication from COE, Pune in 2008. He received PhD degree from Dr C V Raman University, Bilaspur in 2017. He has authored/co-authored 77 research papers in various reputed international publishers' journals/conferences, such as IEEE and Springer. His research interest includes Video Shot Boundary Detection, Design of IoT based applications and drone applications. He has 441 citations, and a h-index of 10 and i-10 index of 10. He guided 24 MTech scholars and 8 PhD scholars are under his supervision. He received MHRD Scholarship during the course of his MTech at COE, Pune.

Anupama Mohabansi

Ms Anupama Mohabansi received a BE in Electronics from Ramdeobaba Kamla Neharu Engineering Collage Nagpur (RTMNU) in 2007, and an ME in Wireless Communications and Computing from G H Raisoni Collage of Engineering for Womens (RTMNU) in 2015. She is pursuing PhD from G H Raisoni Institute of Engineering and Technology Nagpur in wireless communication. Her chapter 2 is published in Book Cognitive Sensors: Vol 1 and one published patent under the title of The Efficacy of Artificial Intelligence and Internet of Things Models for Soil Monitoring and Crop Management. Recently she is working as Assistant Professor in G H Raisoni Institute of Engineering and Technology Nagpur (GHRIET) with teaching experience of 8 years.

Sheetal Mungale

Ms Sheetal Mungale received her MTech degree in Digital Communication from Rajiv Gandhi Proudyogiki Vishwavidyalaya University in 2014 and BE degree in Electronics Engineering from Rashtrasant Tukadoji Maharaj Nagpur University in 2006. She is working as assistant professor in Electronics and Telecommunication Engineering Department in G H Raisoni College of Engineering, Nagpur. She has a total of 15 years of teaching experience in various engineering colleges. She has three patents and two copyrights to her credit. She has more than eight international journal publications and also conferences to her credit. She has organized two FDP and has attended more than 25 Faculty Development program/ STTP/Workshops. She is a Free Life member in various technical societys, i.e., ACM.

Manjushree Nayak

Dr Manjushree Nayak is currently working as Associate Professor in the Department of Computer Science and Engineering at NIST Institute of Science and Technology (Autonomous), Berhampur, Odisha. Former working as Assistant Professor in MSIT Department of MATS University Raipur (C.G). She has 18 years' experience of work in academia and administration. She has completed her PhD in Computer Application and IT. Her areas of interest are big data, machine learning, soft computing, wireless sensor network, IoT, deep learning, and data analytics. She has published two books. She has published more than 25 papers in national and international journals. Reviewers of many reported journals like IEEE, IGI Global, and Oxford. She has delivered more than 5 keynote/invited talks and chaired many technical sessions in national and international conferences, Life members in Indian Science congress and International Association of Engineers, Computer Science Teacher Association, Science Publishing.

Rajendra B Nerli

Dr Rajendra B Nerli is a distinguished urologist, currently holding multiple esteemed positions including Director of Clinical Services and Director of KLE'S Kidney Foundation, as well as being a Professor in the Department of Urology at JN Medical College, KLE Academy of Higher Education and Research. With an extensive academic background comprising an MS, MCH, PhD, and MBA, he specializes in paediatric urology, female urology, and urooncology. Dr Nerli has made notable contributions to the field, spearheading the renal transplant program in Belagavi and establishing the MCH Training program at J N Medical College. His exceptional achievements have been recognized with prestigious awards, including the Karnataka State Rajyotsava Award, Doctor Joshi Gold Medal, and Pinnamaneni Gold Medal and Oration. He is highly regarded in professional associations such as the Indian Urology, American Urology Association, International Continence Society, and Society of Paediatric Oncology International. Dr Nerli has an impressive publication record, having authored over 350 papers and contributed to 10 book chapters.

Chandrashekhar Himmatrao Patil

Dr Chandrashekhar Himmatrao Patil, PhD, received his MCA from University of Pune and PhD degrees in Computer Science from Bharati Vidyapeeth Deemed University, Pune in 2016. He joined the Department of Computer Science (at present, Dr Vishwanath Karad MIT World Peace University) as a faculty member in 2008. Since December 2008, he has been with the Department of Computer Science and Applications, where he is currently an Associate Professor. His research interests include pattern recognition, image processing, sensing data analysis, and satellite image processing. Dr C H Patil has authored/co-authored seven publications in various high impact factor, peer-reviewed, journals. Dr C H Patil has written two book chapters and he presented about 45 papers in international conferences. Dr C H Patil has granted two patents and three copyrights.

Pushkar Bansidhar Patil

Pushkar Bansidhar Patil is currently working as a Graduate Engineer Trainee at Mercedes Benz Research and Development India. He has successfully completed his BE degree in Electronics and Communication Engineering from K L S Gogte Institute of Technology, Belgavi, Karnataka. Pushkar has gained extensive experience in the field of AI, ML, and DL through his active involvement in multiple projects during his engineering studies. These projects have significantly piqued his interest in the field, motivating him to further explore and contribute to the advancements in AI, ML, and DL.

Anu G Pillai

Anu G Pillai is working as Assistant Professor in Electrical Engineering Department at Kalinga University, Raipur. She got her Diploma in Electrical Engineering from Government Polytechnic College, Durg affliated to Rajiv Gandhi Prodhyogiki Vishwavidyalay, Bhopal, MP in 2004, and completed her BE in Electrical Engineering in 2007 from MPCCET Bhilai followed by MTech (Power System Engineering) in 2014 from SSTC Bhilai. She bagged second position in the university exam list in her BE (final year) followed by honors in MTech. Presently she is pursuing her PhD in Electrical Engineering from Kalinga University, Raipur. She had worked as Assistant Professor in the Electrical Engineering department at SSTC, Bhilai from June 2008 till November 2022. She has published over 20 papers in renowned international and national journals and conferences, including the Scopus journal. She also undergone AICTE Parakh program training and is one of the experts involved in building assessment questions for AICTE. Her area of interest includes renewable energy, grid integration and power quality. She is the IQAC Coordinator of EE Department in Kalinga University and a professional member of the Institute for Engineering Research and Publication (IFERP) and IAENG.

Roshni Rahangdale

Roshni Rahangdale received a BE in Electrical Engineering (EE) from the BIT, Durg in 2002, and an ME (EE) from the SSTC, Bhilai in 2010. She is pursuing a PhD from BIT, Durg. Presently she is working as an Assistant Professor in the EE department in SSTC, Bhilai with a total experience of 20 years. Her research interests are power system protection, digital relaying, power quality and power system transients.

Mainak Sadhya

I am **Mainak Sadhya**, a student currently in my second year at PDPM Indian Institute of Information Technology, Design, and Manufacturing Jabalpur (IIITDMJ). I am pursuing a Bachelor of Technology (BTech) degree in the department of Electronics and Communication Engineering (ECE). I was born on 24/12/2002 and Jabalpur is my hometown. I have completed my schooling from Christ Church Boy's Senior Secondary School which is located in Jabalpur as well.

At IIITDMJ, I have been studying various subjects related to electronics, communication systems, and signal processing which are said to be the pillars of this field. The institute focuses on information technology, design, and manufacturing, providing me with a well-rounded education in these areas.

Being in the second year of my BTech program, I have already had the opportunity to engage in academic pursuits, collaborate with fellow students, and work on hands-on projects. I have been able to explore my interests within the field of electronics and communication engineering.

I am excited about the opportunities ahead as I continue my academic journey. I hope to make significant contributions to the field and leverage my knowledge and skills to positively impact society. I am really excited to see what the future holds for me as I keep working harder and fulfill the ambitions of mine as well as the dreams harbored by my parents.

Overall, I am proud to be a student of PDPM IIITDMJ, pursuing my passion for Electronics and Communication Engineering and looking forward to the future.

Sourabh Sahu

Dr Sourabh Sahu obtained an MTech and PhD in Electronics and Communication Engineering (ECE) from MNIT Jaipur. Instrumental in programming languages like Python (including Tensorflow, Scikit-Learn, and Keras for Machine Learning and Deep Learning algorithms), C and C++, and various numerical computational tools such as OptiFDTD, Lumerical MODE Solver, MATLAB, SCILAB. He is presently working as an Associate Professor in the Department of ECE at GGITS Jabalpur, and before that worked as Assistant Professor at Jabalpur Engineering College, Jabalpur under the TEQIP-III project and Adhoc faculty at NIT Raipur. He is the recipient of a research project under the Collaborative Research Scheme (CRS) of TEQIP-III with a total grant amount of Rs. 19.45 Lacs from MHRD, Government of India. He has authored more than 17 research papers in indexed journals and conferences of international and national repute. He has also served as a reviewer for various journals. His area of interest includes designing, pre-processing, validating, debugging, and modeling artificial intelligence and machine learning algorithms for predictive data analysis.

Amin Sakhaei

Amin Sakhaei, is currently a senior BSc student in Biomedical Engineering at Amirkabir University of Technology, Tehran, Iran. Previously, he gained experience as a research intern at The Institute for Cognitive and Brain Sciences (ICBS) at Shahid Beheshti University, Tehran, Iran. During this period, he worked under the supervision of Dr Reza Khosrowabadi.

Currently, Amin is immersed in research on False Memory and Biomedical Image Processing. He is conducting this research under the supervision of Dr Zahra Ghanbari in her lab. His primary research interests include medical and biomedical image and signal processing, deep learning, cognitive neuroscience, memory, and brain connectivity.

Rishab Sarkar

Rishab Sarkar graduated from the University of Calcutta, West Bengal, India with a BTech degree in Electronics and Communication Engineering. With his undergraduate studies completed in 2022, he is currently pursuing an MTech in Signal Processing at the National Institute of Technology (NIT), Calicut, Kerala, India. His research interests revolve around cutting-edge fields such as computational methods in medical imaging, medical image analysis, computer vision, and machine learning.

Ruhi Uzma Sheikh

Ruhi Uzma Sheikh received her BE in EEE from Pt. Ravi Shankar Shukla University, Raipur in 2005, an ME in Power System Engineering from CSVTU, Bhilai in 2010 and a PhD Degree in EE from Sri Satya Sai University of Tech and Medical Science, Sehore, Bhopal (M.P.) in 2021. She is currently working as Assistant Professor at RTM Nagpur University, Nagpur and has 16 years of teaching experience. She has publications in 30 research papers in various journals and conferences. Her area of interest is real time applications for energy management and smart grid applications.

Richa Ravi Siddannavar

Richa Ravi Siddannavar is a budding engineer with a BE degree in Electronics and Communication Engineering from K L S Gogte Institute of Technology, Belgavi, Karnataka. During the period of engineering studies, she worked on projects involving plant nuclei segmentation, doctor's handwriting recognition, and prostate cancer detection, showcasing her passion for AI, ML, and DL. With a strong interest in these fields and their applications across a wide spectrum, Richa aims to build a successful career in AI, ML, and DL. She possesses proficiency in these technologies and is eager to contribute to cutting-edge research and innovation.

G Sreelekha

Dr G Sreelekha received her BTech degree in Electronics and Communication Engineering from the College of Engineering Trivandrum, Kerala, India, in 1997, her MTech degree in Digital Electronics from Cochin University of Science and Technology, Kerala, India, in 2000 and her PhD from National Institute of Technology Calicut, Kerala, India, in 2009. She joined as a Lecturer in the Department of Electronics and Communication Engineering, National Institute of Technology Calicut in 2000 and is currently working as a Professor in the same department. Her major research interests are image and video compression, biomedical signal processing, efficient architectures for DSP and machine learning. Dr Sreelekha has authored/co authored 14 publications in various high impact factor, peer-reviewed journals and has 28 publications in various international conferences. She is a senior member of the IEEE, the Institute of Electronics, Information and Communication Engineers.

P V Sudeep

Dr P V Sudeep is currently working as Asst. Professor in the Department of Electronics and Communication Engg., National Institute of Technology (NIT), Calicut, Kerala, India. He received his PhD from NIT, Tiruchirappalli and MTech and BTech from the University of Kerala, India. Research interests lie in the domains of computational methods in medical imaging, medical image analysis, computer vision, NLP and machine learning. He has published several research articles in various high impact factor, peer-reviewed international journals and conferences. He is also a member of various professional organizations such as IEEE and ISTE. He is also serving as a reviewer and an editorial board member for various reputed peer reviewed journals and technical committee member for various international conferences.

Saurabh Tewari

Dr Saurabh Tewari received a BTech in Electrical and Electronics Engineering from Uttar Pradesh Technical University in 2012. He qualified for the GATE examination in 2013 and joined the MTech Petroleum engineering program at Rajiv Gandhi Institute of Petroleum Technology (RGIPT). He completed an MTech degree (2015) in Petroleum Engineering and PhD (2021) from RGIPT. his industrial experience includes 8 months as an HSE officer in a naval dockyard. He also served Algo8 AI Private Limited (Canada based company) during 2021. He is presently working as Associate Processor in the Computer Science Department at Gyan Ganga group of Institutions, Jabalpur. He is an expert in AI and data-driven models. He was featured in the Departmental magazine of the European Association of Geoscientists and Engineers (EAGE), Suez University, Egypt, 2019, as a Young Petroleum Researcher. He also won first prize in the South Asia Pacific Regional Paper Contest 2019, organized by the Society of Petroleum Engineers (SPE) and was selected for representing South Asia Region in International Paper Contest, ATCE, 2019, USA. He was nominated for Young Professional of The Year 2018–19, SPE ADIPEC, 2019. He also received a Gold medal from 'JIM CORBET' for attaining All India Rank One in an environmental awareness competition in 2006. In his credit there are six papers in international journals and one in a national journal.

Shruti Tiwari

Dr Shruti Tiwari received her BE from Pt. RSSU, Raipur, ME and PhD degrees from CSVTU, Bhilai in EEE and Electrical Engineering in 2007, 2009, and 2021, respectively. She is currently working as a Sr Assistant Professor in the Faculty of Engineering, Shri Shankaracharya Technical Campus, Bhilai from July 2007. She has 15 years of experience. Her research interest includes energy management systems, sustainable energy, and signal processing. She authored/co-authored various publications in various peer-reviewed journals. She has got her work patented. She has written 1 book chapter and 1 book and presented papers in national/international conferences.

Prajakta Upadhye

Ms Prajakta Upadhye received her MTech degree in Electronics (Communication) from Rashtrasant Tukadoji Maharaj Nagpur University in 2014 and a BE degree in Electronics and Telecommunication from Amravati University in 2003. She is currently working as Assistant Professor in the Electronics and Telecommunication Department in GH Raisoni Institute of Engineering and Technology Nagpur and has 12 years working experience in total. Her research interests include image processing, and IoT application. Prajakta Upadhye has authored/co-authored three publications in various journals and conferences. She has written one book chapter and had one patent to her credit. She is member of the Indian Society for Technical Education, Institution of Engineers (India), and the International Association of Engineers professional societies.

Amol D Vibhute

Dr Amol D Vibhute received his PhD in Computer Science under the domain of Geospatial Technology, MPhil and MSc in Computer Science from the Department of Computer Science and Information Technology, Dr Babasaheb Ambedkar Marathwada University, Aurangabad, MH-India. He is an Assistant Professor at the Symbiosis Institute of Computer Studies and Research (SICSR), Symbiosis International (Deemed University), Pune, Maharashtra, India. He has more than 8.5+ years of academic experience in research, innovation, and UG/PG level teaching. His six Indian patents are at the granting stage, one international patent has been granted, and he authored/co-authored over 65+ referred journals, book chapters, and conference papers in reputed international journals and conferences indexed by Scopus/SCIE/UGC. The current publication citations of his research are more than 600+, 242, and 104, with an h-index of 13, 7, and 5 as per Google Scholar, Scopus, and Web of Science, respectively. His specialization is hyperspectral-multispectral remote sensing, GIS, and remote sensing of soil, crops, urban areas, etc. His research interests include geospatial technology, digital image processing, pattern recognition, big data analysis, the Internet of Things (IoT), machine learning, and artificial intelligence. He is an ISPRS, IAEng, CSTA, IACSIT, and ISCA member.

IOP Publishing

Image Processing with Python
A practical approach
Irshad Ahmad Ansari and Varun Bajaj

Chapter 1

Basics of image analysis and manipulation using Python

Akanksha Dixit

A digital image is analyzed and manipulated during image processing with the main objective of improving its quality or acquiring data for later use. Analyzing and manipulating an image with various computer-based algorithms is simply called image processing. It includes a variety of features, including image storage, information extraction, representation, enhancement, and image interpretation, etc. For such types of image tasks, Python becomes an acceptable choice. This chapter provides a fundamental overview of each of these various facets of image processing in addition to a primer on practical image processing using Python packages. Here, all the code samples are written in Python 3. In this chapter different Python libraries are introduced, and Python codes are written to analyze the image using these libraries.

1.1 Introduction

Images are virtually everywhere. Humans can easily recognize the images or objects within seconds, but for computers, it is a taught task to identify the image in a similar way. A digital image is viewed as matrices of 0s and 1s by a computer [1, 2]. The smallest component in an image is a pixel. A digital image is created as a collection of pixels. The number of channels in each pixel varies. When a picture is colored, it has three channels: red, green, and blue, as opposed to one channel when it is monochrome [3].

Digital images are made up of pixels; these pixels are represented in 2D as f(hr, vr), which is the amplitude or intensity of the pixels at that location (hr,vr) [4]. If a matrix defines an image, then hr represents the row, vr represents the column as shown in figure 1.1, and f(hr,vr) is the location of that pixel, giving some value called intensity. The function f(hr,vr) depends on the illumination a(hr, vr) and reflectance b(hr,vr).

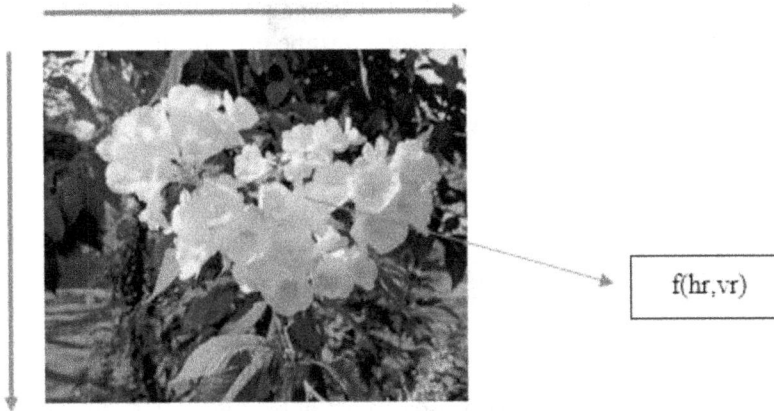

Figure 1.1. A sample image.

f(hr, vr) = a(hr, vr). b(hr, vr)

This chapter defines an image, and how these images are processed using the Python programming language. There are many Python libraries [5] available for image processing. In this chapter, some of them were used such as OpenCV, `matplotlib` and `NumPy`.

Before delving into the analysis and manipulation of an image, it is crucial to understand what image processing is and how it fits into the larger scheme of things. The most common name for image processing is 'digital image processing,' and the field in which it is most usually applied is 'computer vision'. Computer vision (CV) and image processing methods both require an input image; in contrast, the output of computer vision can be certain features or information about the image, whereas the output of image processing is also an image.

1.2 Image analysis and manipulation

Image analysis involves dissecting an image into its individual components to obtain usable information from it. The tasks that can be incorporated in image analysis include identifying shapes, recognizing edges, removing noise, counting objects, computing statistics for texture analysis, and determining image quality.

The term 'image analysis' is broad and incorporates a variety of methods, most of which belong to one of the categories that follow: image enhancement, image segmentation, region analysis and noise removal.

Image manipulation is a process that is applied on an image to get a desired output. The image needs to be modified or altered to get desired output. Some image manipulation operations include image cropping, image resizing, image filtering, etc. To increase or decrease the size of an image, an image resizing function is used. Image cropping is used to select any desired area from an image.

The following steps are mainly involved in image processing (figure 1.2):
1. **Input acquisition/storage:** The image must be taken using a camera and saved on a device (hard disk). The image should be saved as a file with a file format like .jpeg, .png, etc.
2. **Load into memory and save to disk:** A data structure must be used to read the image from the disc and save it in memory. Here the numpy ndarray data structure is used to arrange the image data.
3. **Manipulation, enhancement, and restoration:** There are many preprocessing algorithms that are applied to the image. For example, enhancing the quality of image filtering is used. A few transforms are applied to the image for sampling and manipulation purposes. Sometimes images need to be restored for noise degradation.
4. **Segmentation:** it is a process used to extract some interesting or useful area, called objects, from an image.

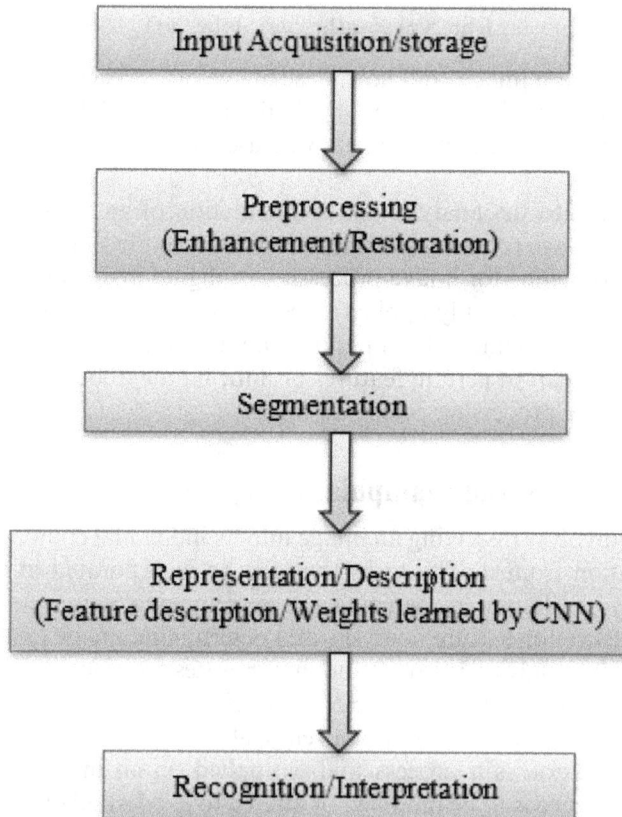

Figure 1.2. Different steps in image processing.

5. **Information extraction and representation:** Images are represented in some alternative forms, for example some hand-crafted features are computed from an image for image processing. With deep learning some features (weights and biases) are automatically calculated in hidden layers of the neural network.

6. **Image understanding and interpretation:** Images can be better understood with the help of some processes such as image classification (for example whether the image is of a cat or a dog) and object recognition (for example finding the location of car objects in each image).

1.2.1 Applications

- **Image correction, sharpening, and resolution correction:** Modern technology makes it feasible to enhance the quality of antique photos. These techniques include sharpening, zooming, edge detection, and high dynamic range editing. All these actions help to enhance the image.

- **Filters on editing apps and social media:** These days, filters are available on many editing and social media apps. The image becomes more visually attractive by using these filters. Filters are typically a collection of operations that alter the colors and other elements of an image to give it a unique appearance. Filters are an interesting application of image processing.

- **Medical technology:** Image processing is used in the medical industry for a variety of tasks including PET scanning, x-ray imaging, medical CT scanning, UV imaging, cancer cell image processing, and much more. The introduction of image processing to the field of medical technology has greatly improved the diagnostics procedure. We can use an illustration from a certain source [10] and can observe the processed image, which is far better, can be used for better diagnostics.

- **Computer/machine vision:** Computer vision is one of the most fascinating and practical uses of image processing. Making a computer see, recognize objects, and process the entire environment is done using computer vision. Computer vision is crucial for self-driving cars, drones, and other devices. Obstacle detection, path recognition, and environmental comprehension are all made easier by using CV.

1.3 Python—features, application

1.3.1 Python

Python is one of the most adaptable and dynamic programming languages today. It is a robust, popular, object-oriented interpreted, and high-level programming language. This enables the development of both straightforward and intricate processes. Additionally, it supports a wide range of programming languages, including Java++ and JSON. Many businesses use it as a backend programming language.

1.3.1.1 Python features

Python offers several practical characteristics that set it apart from other programming languages and make it popular. It offers dynamic memory allocation, allows procedural programming, and is object-oriented. There are some essential features discussed below.

1. **Easy to learn and use**

   Python is a beginner-friendly language. It is easy to use and learn compared to other programming languages. Also, it is easy to code advanced algorithms including advanced AI, machine learning, and data visualization, in Python, even for beginners.

2. **Cross-platform language**

   Python can be installed on any system such as Windows, Linux, UNIX, and Macintosh, etc. So, it can be concluded that Python is a portable language. And it's very easy to start with Python.

3. **Free and open source**

   Python is free and very accessible. It can be downloaded from its official website www.python.org.

4. **Large standard library**

   It has many standard libraries, and this is one of the primary reasons Python is growing very fast. There are many ready to use libraries (e.g., Opencv, Scikit-image, Matplotlib, etc.) available for the various fields such as machine learning and web development.

1.3.2 Python applications

- Desktop GUI
- Audio and video application
- Image preprocessing
- Web development
- Game development
- Data science and data visualization
- CAD application
- Machine learning and artificial intelligence
- Business applications

1.3.3 Python libraries

1. OpenCV
2. Numpy
3. Matplotlib
4. Scikit-Image
5. Scipy
6. Pillow/PIL
7. SimpleITK
8. Mahotas

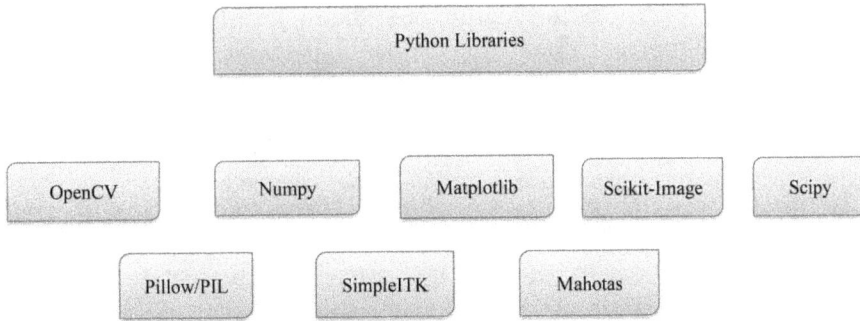

Figure 1.3. Python library.

The Matplotlib module in Python can be used to create 2D graphs and plots. The Python package NumPy is used to store an image in the form of array and NumPy ndarray data structure is used to arrange the image data. The industry-leading machine learning (ML) library is open source and called Scikit-learn. It is used to create machine learning model for image processing. Numerous functions from NumPy and data science have been improved and added to `SciPy`. `OpenCV`, Mahotas and Scikit-image libraries are used for different image processing algorithms (figure 1.3).

These are some commands [7] which are used to install Python libraries with pip (standard package manager).PIP is a tool, which is used to install the Python packages and dependencies.

```
>>  pip install opencv-python.
>>  pip install numpy.
>>  pip install scipy.
>>  pip install scikit-image.
>>  pip install matplotlib.
>>  pip install pillow.
>>  pip install simpleITK
```

1.4 OpenCV package

OpenCV is an open-source library package with real-time computer vision capabilities. Applications involving machine learning and image processing use it. Python is among the many programming languages that it supports. It is free for academic and commercial purposes due to its BSD license (Berkeley source distribution).

There are many platforms to program in Python [7], such as Jupyter Notebook, Python Shell, PyCharm and Visual Studio Code. Jupyter Notebook is a widely used interactive computing platform. For installing Anaconda Navigator—Jupyter Notebook, please refer to:

https://docs.anaconda.com/free/anaconda/install/windows/
https://docs.jupyter.org/en/latest/install/notebook-classic.html

1.4.1 OpenCV installation

```
pip install opencv-python
```

Run this command on Anaconda Navigator—Jupyter Notebook.

1.4.2 Importing the module from package

Python packages contain many modules, whereas Python modules may contain multiple classes, functions, variables, etc. Simply, a Python package is a folder that includes different modules as files. In Python, a package is a group of modules that include pre-written programs. These packages enable us to import modules fully or partially. Importing module works like this:

```
➤ import cv2
➤ from PIL import Image
➤ import numpy as np
➤ from matplotlib import pyplot as plt
➤ from matplotlib import image as mpimg
➤ from skimage import io
➤ from warnings import filterwarnings
➤ import tensorflow as tf
➤ from tensorflow import io
➤ from tensorflow import image
```

1.4.3 Reading and displaying image with different libraries

Color images, grayscale images, binary images, and multispectral images are the different categories that digital images might fall into. Each pixel's color information is present in a color image. Grayscale images are images with only shades of gray, whereas binary images have only two colors, black and white. Photographs that capture image data from various electromagnetic spectrum wavelengths are referred to as multispectral images. Consider the image in figure 1.4, which is an image of a flower. The image is read from a local directory.

- Using OPenCV library

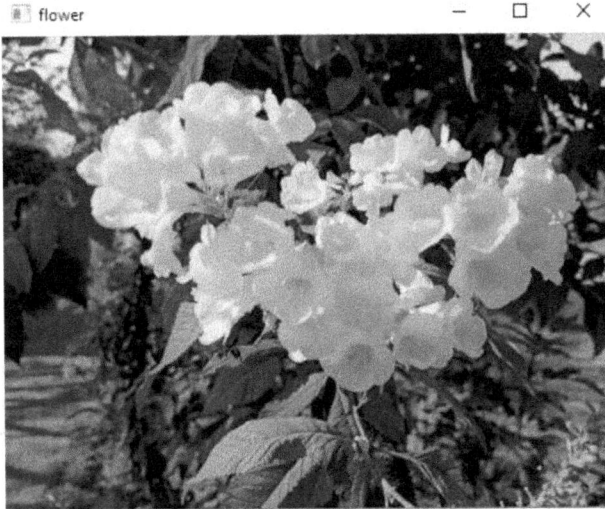

Figure 1.4. Flower image using cv2.

```
# cv2.imread(image_path_with_file_extension, flag)

# cv2.imshow(image title, image path variable)

# Importing libraries
import cv2
import numpy as np
# Reading the image
image = cv2.imread ("3n.jpg", -1)
# Displaying the image
cv2.imshow('flower', image)
# Wait till key is pressed
cv2.waitKey (0)
# Close all windows

cv2.destroyAllWindows ()
```

Note: Since the image is already available in the working directory in this case, only the image name and extension type are required. However, if the image is missing, an absolute path from local computer drives or even the internet should be defined.

```
image = cv2.imread('C:\\Users\\sp1014\\Pictures\\3n.jpg', -1)
```

A GUI window is generated to display the image for use on the screen. The title of the GUI window screen, supplied in string format, must be the first parameter. The cv2.imshow() method allows us to display the picture in a pop-up window. However, if the window attempted to close, the user can feel trapped by the window. So, a straightforward 'waitKey' method is employed to combat that. By running code in a new cell:

```
# Displaying the image
cv2.imshow('flower', img)
# Wait till key is pressed
cv2.waitKey(0)
# Close all windows
cv2.destroyAllWindows()
```

The window will remain open until we close it using parameter '0' in the 'waitKey' in this case. Instead of 0 seconds, the time can also be specified in milliseconds, indicating how long it should take to open. To close or remove GUI windows from the screen or memory, use the cv2.destroAllWindows() method.

- Using Matplotlib library

The main purpose of this module is data visualization. However, using plotting techniques, the images can be examined graphically, with each pixel located on a 2D x–y-axis.

```
#Importnig the image using matplotlib library
from matplotlib import image as mpimg
#reading the image
ima = mpimg.imread("3n.jpg")
#Importnig the plot function from matplotlib library
from matplotlib import pyplot as plt
plt.title ("flower")
plt.xlabel ("X_axis")
plt.ylabel ("Y_axis")
#displaying the image using matplotlib library
plt.imshow (ima)
plt.show()
```

Output of above code can be observed in figure 1.5.

If an image is read using cv2 and needs to be displayed using Matplotlib (figure 1.6) then the output is:

```
plt.imshow(ima)
```

OUT: (figure 1.6)
- Using pillow library

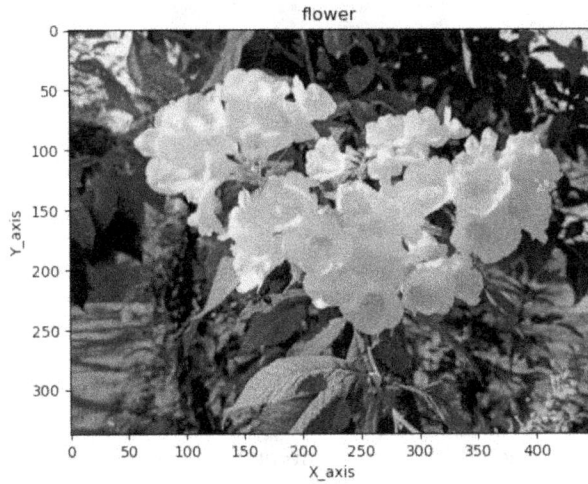

Figure 1.5. Flower image using Matplotlib.

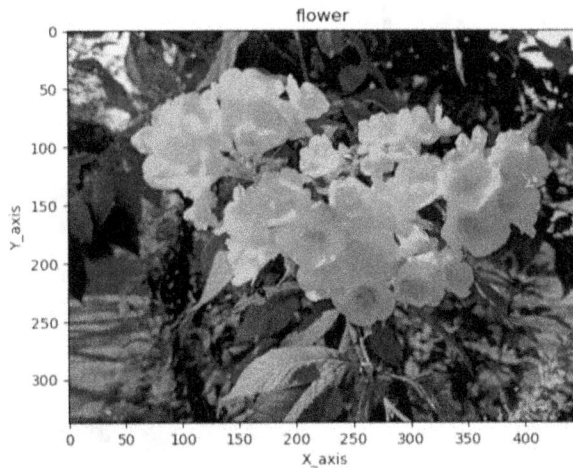

Figure 1.6. Displaying image using Matplotlib.

```
#Importnig image using PIL library
from PIL import Image
#open the image
img = Image.open("flowerr.jpg")
img.show()
```

The image is shown using the built-in photo program of OS. This library frequently provides easy-to-use ways for manipulating images.

OUT: (figure 1.7)

Figure 1.7. Flower image using pillow.

1.4.4 Image properties

Shape: An image always has a shape. The picture's boundaries' length, which includes their height and width, can be thought of as the shape. The following code shows the shape of a flower image which is 512×512 with 3 channels.

```
#Importnig all necessary libraries
import cv2
import numpy as np
#Reading an image
im = cv2.imread('3n.jpg', -1)
#Displaying the shape of an image
print(im.shape)

OUT: (512, 512, 3)
```

The above code is the simplest approach to print the shape of the image, however, to understand it better, the image shape can be extracted by their identity also. In the next code three parameters are considered; height, width, and channel for displaying height of the image, width of the image and number of channels in the image, respectively.

```
height, width, channel = im.shape
print("Dimensions of the image is:", height, "pixels_Width:", width,
"Number of Channels:", channel)

OUT: Dimensions of the image is: 512 pixels_Width: 512 Number of Channels
: 3
```

It always returns three values for the color image, including the image's height, width, and number of channels. The height and width of the image will be returned if grayscale images are processed, in which the shape would be 2, but if only height and width variables are defined in code and channel is not, then in that case there is a value error that reads 'not enough values to unpack' which means the expected variable for function was three but only two variables are defined.

Type: Type of image can also be found using the 'type' method. This approach can be used to learn how image data is represented.

```
#Importnig all necessary libraries
import cv2
import numpy as np
#Reading an image
im = cv2.imread('3n.jpg', -1)
#Displaying the shape and class of the image
print(im.shape)

print(type(im))

OUT (512, 512, 3)
<class 'numpy.ndarray'>
```

Image pixel values: Small samples combine to form an image. Pixels are the name for these samples. If an image is enlarged as much as it can for better comprehension. Image can be seen, which is broken up into many squares. These are the pixels that make up an image when they are all put together.

A matrix is one of the straightforward ways to represent an image. Even better, a matrix is used to produce and preserve an image.

```
#Importnig all necessary libraries
import cv2
import numpy as np
#Reading an image
image = cv2.imread("3n.jpg",-1)
#Showing the image in the form of matrix
print(image)

OUT:
[[105 101 106 ...  57  56  55]
 [109 105 110 ...  56  55  54]
 [115 112 121 ...  55  55  53]
 ...
 [127 122 116 ...  95  97 100]
 [131 123 115 ...  99 100 103]
 [131 122 112 ... 102 103 105]]
```

1.4.5 Image resolution:

The number of pixels in an image could be referred to as its resolution. When there are more pixels in an image, the image quality improves. The image's shape indicates how many rows and columns are present in an image. An image's resolution determines how much detail it has. More visual detail translates to 'higher resolution'.

By multiplying the number of columns by the number of rows, it is possible to get the total number of pixels in the image. An image with a size of 320 by 240 pixels, for

Figure 1.8. Image resolution 512×512.

Figure 1.9. Image in grayscale mode.

instance, has a total of 76 800 pixels. Image can be seen in figure 1.8 with the resolution of 512×512 and in gray scale (figure 1.9).

```
image = cv2.resize(img, (512, 512))
```

1.4.6 Opening in grayscale mode

```
Import cv2
img =
cv2.imread('C:\\Users\\sp1014\\Pictures\\3n.jpg',cv2.IMREAD_GRAYSCALE)
```

Image can be seen in the grayscale mode in figure 1.9.

1.4.7 Save image

It happens rather frequently while working with images in image processing apps that you need to save the final resulting image or keep intermediate outcomes of image transformations. NumPy ndarray is used to store the images when using OpenCV Python. Use the OpenCV Python library's, cv2.imwrite() function to store an image to the local file system.

The syntax for writing the image in specified file location is given by

```
cv2.imwrite('/path/to/destination/image.png', image)

#Importnig all necessary libraries
import cv2
import numpy as np
#Reading and Displaying an image
image = cv2.imread("3n.jpg", -1)
cv2.imshow('flower_Image', image)
#Wait till key is pressed
cv2.waitKey(0)
#Writing the image
cv2.imwrite("saved_flower_Image.jpg", image)
```

Note: After running the above code, please close the image window and only then the true response will be displayed in the screen.

The fact that cv2.imwrite() returned true indicates that the file was successfully written to the specified path. If the path is not specified then the image is saved into the current file directory. Reading the imwrite() return value is crucial because there may occasionally be several factors that cause the disc write operation to fail and leave the image off the disc.

OUT: saved_flower_Image.jpg

1.5 Some Python code and explanation

1.5.1 Image sharpening

Knowing that kernels are used for various image-processing tasks [6], such as picture blurring, image sharpening, embossing, edge detection, and more, is beneficial for the new researchers of the domain. There are some common types of kernel, which are as follows:

- Convolution Matrix.
- Mask.
- Matrix/Array.

For the operations of blurring, sharpening, embossing, edge detection, and others, a kernel must be applied to the image pixels.

To perform operations like blurring, sharpening, embossing, edge detection, etc., kernel functions need to be convolved with the test image. There is a general formula for convolution (blurring, sharpening, etc.) shown as equation (1.1).

$$g(x, y) = \omega * f(x, y) = \sum_{dx=-a}^{a} \sum_{dy=-b}^{b} w(dx, dy) f(x + dx, y + dy) \qquad (1.1)$$

where $f(x, y)$ is the original image, $g(x, y)$ is the filtered image, and ω is the filter kernel. Every component of the filter kernel is regarded as having the values - as $-a \leqslant dx \leqslant a$ and $-b \leqslant dy \leqslant b$.

Kernels are as follows: (figure 1.10)

Python requires a filter2D () function to sharpen an image. In this function there are three important parameters that should be passed, which are as follows:

- *src*: This is the image that will be sharpened, or the source image.

Operation	Kernel ω	Result $g(x, y)$
Identity of Image	[0 0 0 0 1 0 0 0 0]	
Edge detection of Image	[1 0 -1 0 0 0 -1 0 1]	
	[0 -1 0 -1 4 -1 0 -1 0]	
	[-1 -1 -1 -1 8 -1 -1 -1 -1]	
Sharpening of Image	[0 -1 0 -1 5 -1 0 -1 0]	

Figure 1.10. Image sharpening kernels.

- *ddepth*: This integer number represents the output image's anticipated depth. This number will be set to be equal to −1. By doing this, the compiler understands that the output image will have the same depth as the input image.
- *kernel*: This is the kernel that will be used to process the image. For sharpening the image, the following kernel is required.

$$\text{Sharphen\_kernel} = \begin{bmatrix} 0 & -1 & 0 \\ -1 & 5 & -1 \\ 0 & -1 & 0 \end{bmatrix}$$

Image sharpening Python code

```
# importing all necessary libraries

import cv2
import numpy as np
from matplotlib import pyplot as plt

# plot the figure
fig = plt.figure (figsize=(15, 8))
# set values for plot variables
row = 2
column = 2
# load the image into system memory
imagee = cv2.imread("3n.jpg",-1)
# image showing window
cv2.imshow ('flower',image)
# waits till any key is pressed

cv2.waitKey (0)
# Close all windows

cv2.destroyAllWindows ()
# sharpening the image using differnt kernels
sharphen_kernel1 = np.array([[0, -1, 0],
            [-1, 5,-1],
            [0, -1, 0]])
edgedetection_kernel1 = np.array([[0, -1, 0],
            [-1, 4,-1],
            [0, -1, 0]])
edgedetection_kernel2 = np.array([[-1, -1, -1],
```

```
                [-1,  8,-1],
                [-1, -1, -1]])
identity_kernel1 = np.array([[0, 0, 0],
              [0, 1,0],
              [0, 0, 0]])
sharphen_image1 = cv2.filter2D(src=imagee, ddepth=-1, kernel=sharphen_kernel1)
edgedetected_image1 = cv2.filter2D(src=imagee, ddepth=-1,
kernel=edgedetection_kernel1)
edgedetected_image2 = cv2.filter2D(src=imagee, ddepth=-1,
kernel=edgedetection_kernel2)
identity_image1 = cv2.filter2D(src=imagee, ddepth=-1, kernel=identity_kernel1)

# Adds a subplot at the 1st position
fig.add_subplot (row, column, 1)

# displaying sharphen image
plt.imshow (sharphen_image1)

plt.title ("sharphen_image")

# for 2nd position subplot
fig.add_subplot (row, column, 2)

# displaying edge detection1 image
plt.imshow (edgedetected_image1)

plt.title ("edgedetected_image1")

# for 3th position subplot
fig.add_subplot (row, column, 3)

# displaying edge detection2 image
plt.imshow (edgedetected_image2)

plt.title ("edgedetected_image2")

# for 4th position subplot
fig.add_subplot (row, column, 4)

# displaying identity image
plt.imshow (identity_image1)

plt.title ("identity_image")
```

The output of the above code can be seen in figure 1.11.

1.5.2 Image resizing

Scaling an image is the process of image resizing. For many machine learning and image processing applications, scaling is helpful [8]. It assists in minimizing the quantity of pixels from an image and that has several advantages, e.g., it can take

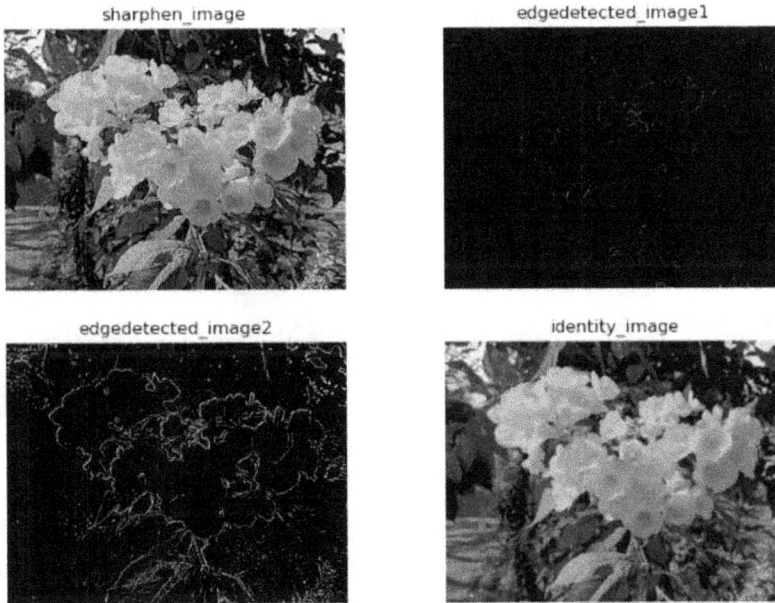

Figure 1.11. Image sharpening.

less time to train a neural network because an image's pixel density increases the number of input nodes, which in turn raises the complexity of the model. Additionally, it facilitates image zooming. Images need to be scaled up or down to achieve the appropriate size. For scaling an image, OpenCV offers us numerous interpolation techniques.

Interpolation techniques for resizing:
- cv2.INTER_AREA: when image needs to be reduced, this method is used.
- cv2.INTER_CUBIC: Although slow, this is more effective.
- cv2.INTER_LINEAR: When zooming is needed, this is mainly used. In OpenCV, this interpolation method is used by default.

Syntax: cv2.resize(source, dsize, dest, fx, fy, interpolation)

Image resizing parameters:
- source: This is the input image array that could be 8-bit, floating-point, or single-channel.
- dsize: This is the size of the output array.
- dest: This defines output array.
- fx: This defines the horizontal axis's scale factor.
- fy: This defines the vertical axis's scale factor.
- interpolation: This may be any one of the above interpolation methods.

Image resizing Python code

```
# importing all necessary libraries

import cv2
import numpy as np
import matplotlib.pyplot as plt
# reading and displaying an image

ima = cv2.imread ("3n.jpg", -1)

cv2.imshow ('flower', ima)

# waits till any key is pressed
cv2.waitKey(0)
# Close all windows

cv2.destroyAllWindows()
# Image Resize operations

half_image = cv2.resize(ima, (256, 256), fx = 0.5, fy = 0.5)
bigger_image = cv2.resize(ima, (1000, 2000))

stretch_image = cv2.resize(ima, (790, 520),
                 interpolation = cv2.INTER_LINEAR)
# give title for resized image

Image_Titles = ["OriginalImage", "HalfImage", "BiggerImage",
"strechedImage "]
Resized_images =[ima, half_image, bigger_image, stretch_image]
Imagecount = 4
 # Displaying all Resized images on a single window

for i in range (Imagecount):
    plt.subplot (2, 2, i + 1)
    plt.title (Image_Titles[i])
    plt.imshow(Resized_images[i])

plt.show()
```

The output of the above code can be seen in figure 1.12.

1.5.3 Image blurring

Blurring an image means reducing its clarity or clarity of detail. Different low pass filter kernels are used to accomplish this [8].

Advantages of blurring:

- It helps to reduce noise. We can reduce noise by employing a low pass filter kernel since noise is thought of as a high pass signal.
- It aids in bringing the image into focus.
- Edges with low intensity are eliminated.
- It assists in concealing the specifics when necessary. For example, when the victim's face is being purposely covered by the police, it is vital to obfuscate.

Figure 1.12. Image resizing.

Types of filters for blurring an Image:

- Gaussian blurring: An image is blurred using a Gaussian function, which results in a Gaussian blur. It is a tool that is frequently found in graphics software and is usually applied to lessen visual detail and noise. Prior to implementing our deep learning or machine learning models, it is also employed as a preprocessing stage.

- Example of a Gaussian kernel (3×3) **1/16** $\begin{bmatrix} 1 & 2 & 1 \\ 2 & 4 & 2 \\ 1 & 2 & 1 \end{bmatrix}$

- Median blur: The median filter is a popular method for removing noise from an image or signal that uses non-linear digital filtering. In digital image processing, median filtering is highly common since it occasionally preserves edges while lowering noise. One of the finest algorithms for eliminating salt and pepper noise is this one.

- Bilateral blur: A non-linear, noise-reduced, edge-preserving image smoothing filter is known as a bilateral filter. It substitutes a weighted average of the intensity values from adjacent pixels for each pixel's intensity. This weight might be based on a Gaussian distribution. As a result, the sharp edges are maintained, and the powerful edges are discarded.

Image blurring Python code

```python
# importing all necessary libraries
import cv2
import numpy as np
# reading and displaying an image
image = cv2.imread("3n.jpg",-1)
cv2.imshow('Original_Image', image)

# waits till any key is pressed
cv2.waitKey(0)

# using a filter for Gaussian blurring
Gaussian = cv2.GaussianBlur(image, (7, 7), 0)
# displaying image with gaussian blurring
cv2.imshow('Gaussian_Blurring_Image', Gaussian)
# waits till any key is pressed
cv2.waitKey(0)

# using a filter for median blurring
median = cv2.medianBlur(image, 5)
# displaying image with median blurring

cv2.imshow('Median_Blurring_Image', median)


# waits till any key is pressed
cv2.waitKey(0)

# using a filter for Bilateral blurring
bilateral = cv2.bilateralFilter(image, 9, 75, 75)
# displaying image with bilateral blurring

cv2.imshow('Bilateral_Blurring_Image', bilateral)
# waits till any key is pressed
cv2.waitKey(0)
# Close all windows
cv2.destroyAllWindows()
#saving all filtered images
cv2.imwrite("OriginalImage.jpg", image)
cv2.imwrite("GaussianBlurringImage.jpg", Gaussian)
cv2.imwrite("MedianBlurringImage.jpg", median)
cv2.imwrite("BilateralBlurringImage.jpg", bilateral)
```

The output of the above code can be seen in figure 1.13.

Figure 1.13. Image blurring 1. original image 2. Gaussian blurring 3. median blurring 4. bilateral blurring.

1.5.4 Image watermarking

Data security can be achieved via a variety of techniques, such as access control, hashing, and cryptography. One of the methods that is used is watermarking. Using the digital watermarking technique, one can embed covert copyright disclaimers or other verification messages into digital audio, video, or image signals and documents [9]. One of the current techniques used to watermark an image is LSB (Least Significant Bit), as it is straightforward and precise. Digital images allow for the direct insertion of information into every pixel or the careful calculation of the busier sections of an image to conceal such messages in less noticeable locations (figure 1.14).

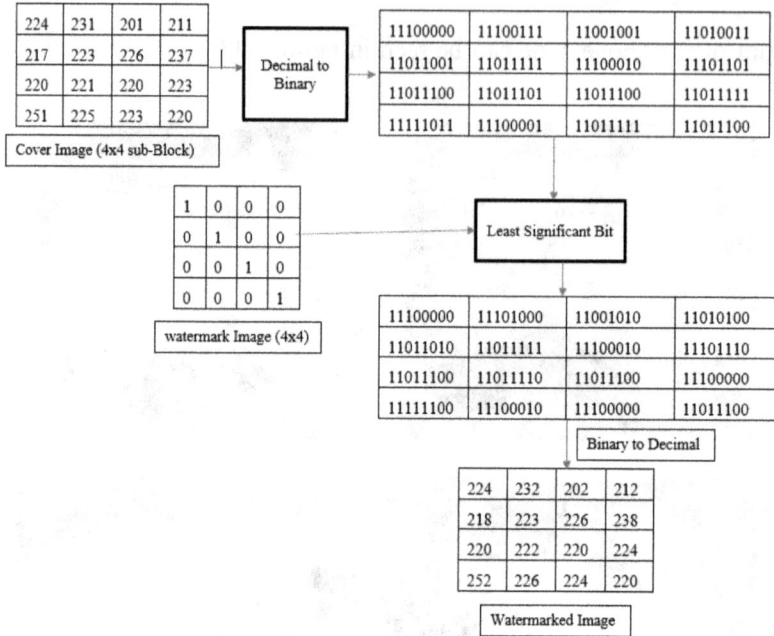

Figure 1.14. LSB watermarking process.

Image watermarking Python code

```
#Importnig all necessary libraries

from PIL import Image
import matplotlib.pyplot as plt
import numpy as np
%matplotlib inline


# Image conversion into binary form using binarize function according to
threshold
def binarize (image_to_transform, threshold):
    # Converting to grayscale
    output_image = image_to_transform.convert("L")
    for x in range(output_image.width):
        for y in range(output_image.height):
            if output_image.getpixel((x, y)) < threshold:
                output_image.putpixel((x, y), 0)
            else:
                output_image.putpixel((x, y), 255)
    return output_image
```

```
#Importnig all necessary libraries

from PIL import Image
import matplotlib.pyplot as plt
import numpy as np
%matplotlib inline


# Image conversion into binary form using binarize function according to
threshold
def binarize (image_to_transform, threshold):
    # Converting to grayscale
    output_image = image_to_transform.convert("L")
    for x in range(output_image.width):
        for y in range(output_image.height):
            if output_image.getpixel((x, y)) < threshold:
                output_image.putpixel((x, y), 0)
            else:
                output_image.putpixel((x, y), 255)
    return output_image


# transformation into binary form (Pixels with values below 128 will be
changed to 0[black] and those with values above 128 to 255[white])
Watermark_Img =binarize(Watermark_Img,128)
plt.imshow(Watermark_Img,cmap='gray')
```

<matplotlib.image.AxesImage at 0x2302d4dcd60>

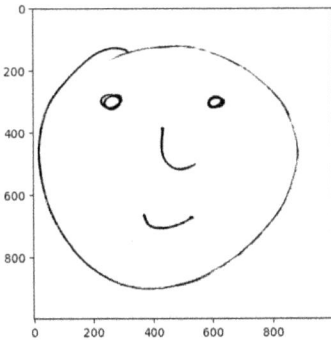

```
# Flatten cover image and convert it into numpy array
Cover_array = np.array(list(Cover_img.getdata()))
#Flatten watermark image and convert it into numpy array
Water_array = np.array(list(Watermark_Img.getdata()))
```

```python
# Watermarking the original image by adding a value of 0 or 1 depending on
the watermarking image's pixel intensity
for p in range(1000000):
    bin1=bin(Cover_array [p])[2:-1]
    x=bin(Water_array [p])[2]
    bin1 += x
    Cover_array [p]=int(bin1, 2)
```

```python
# Converting watermarked array into image
Cover_array =Cover_array.reshape(1000,1000)
enc_img = Image.fromarray(Cover_array.astype('uint8'), Cover_img.mode)
plt.imshow(enc_img,cmap='gray')
```

```python
# Extracting watermark from watermarked image
Ext_array = np.array (list(enc_img.getdata()))
for p in range(1000000):
    if bin (Ext_array [p])[-1]=='1': Ext_array [p]=255
    else: Ext_array [p]=0
```

```python
# Display extracted watermark
Ext_array =Ext_array.reshape(1000,1000)
enc_img1 = Image.fromarray(Ext_array.astype('uint8'), Watermark_Img.mode)
plt.imshow(enc_img1,cmap='gray')
```

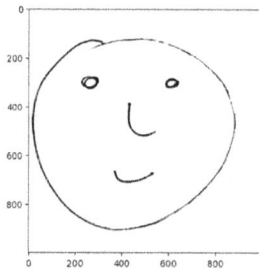

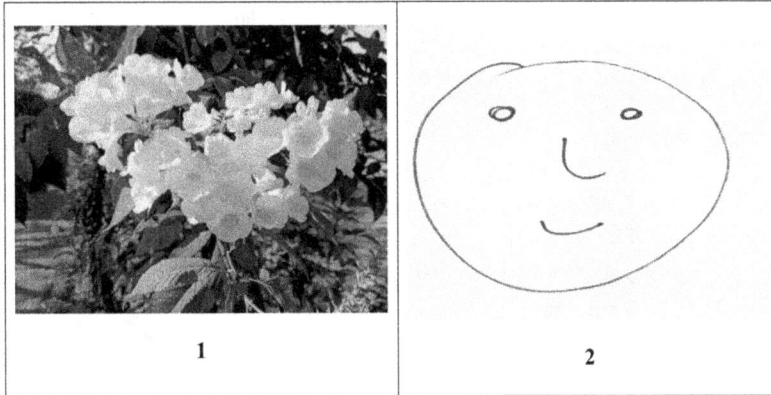

Figure 1.15. Original images 1. Cover image 2. Watermark image.

The cover image and watermark image used in the above code can be observed in figure 1.15.

1.6 Conclusion

In this chapter, a basic introduction of images and different aspects regarding analysis and manipulation of images were introduced. The most widely used Python image processing package, OpenCV, with different Python libraries was then covered. With these libraries images were analyzed via properties like image size, shape, and class, and with some basic operations like reading, displaying writing and saving the image. Finally, some image manipulation operations like image resizing, image blurring, image sharpening and image watermarking were discussed with their Python code. Please keep in mind that this chapter only touches on the surface of image processing (analysis and manipulation), and that there is much more to learn about this field than can be contained in a single chapter.

References

[1] Awcock G J and Thomas R 1995 *Applied Image Processing* (Basingstoke: Macmillan) pp 111–8
[2] Petrou M M and Petrou C 2010 *Image Processing: The Fundamentals* (New York: Wiley)
[3] Koschan A and Abidi M 2008 *Digital Color Image Processing* (New York: Wiley)
[4] Russ J C 2006 *The Image Processing Handbook* (Boca Raton, FL: CRC Press)
[5] https://packaging.python.org/tutorials/installing-packages/
[6] Russo F 2002 *An image enhancement technique combining sharpening and noise reduction IEEE Trans. Instrum. Meas.* **51** 824–8

[7] Jungo A and Scheidegger O 2021 *A Python package for data handling and evaluation in deep learning-based medical image analysis Comput. Methods Prog. Biomed.* **198** 105796

[8] Kinser J M 2014 *Image Operators: Image Processing in Python* (Boca Raton, FL: CRC Press)

[9] Dey S 2018 *Hands-On Image Processing with Python: Expert Techniques for Advanced Image Analysis and Effective* (Birmingham: Packt Publishing)

[10] López A F J 2016 *Teaching image processing in engineering using Python IEEE Rev. Iberoam. Tecnol. Aprendiz.* **11** 129–36

Chapter 2

Digital image processing using Python language

**Devanand Bhonsle, Ravi Mishra, Swati Hadke, Anupama Mohabansi,
Prajakta Upadhye and Sheetal Mungale**

In this chapter a methodological study on image processing has been done, which may be used in various applications. The quality of output image must always be enhanced version of input image. The quality of the image degrades due to various reasons, in which introduction of noise signal plays a significant role. During the acquisition of an image some unwanted signals may be introduced in it, which deteriorate the quality of the image. Hence it is required to remove these noise signals. However, it is challenging to eliminate the interference from noise signals entirely. Nevertheless, their impact can be reduced to a degree that is deemed acceptable for various applications. To remove these noise signals it is desired to study the characteristics of different noise signals, which may help the researchers to make a decision about the selection of a filter for a particular noise signal. However, applying the filtering technique may lead to the suppression of important information like edges, ridges, contours and other fine details which may be helpful for various applications of image processing. Therefore, the goal is to decrease the noise signals without sacrificing the fine details of the images. Image denoising is a preprocessing task. This image is now ready for other operations such as image segmentation, edge detection, object recognition, image sharpening, classification, feature detection and matching, analysis and manipulation of image, etc. To perform all the aforementioned operations. MATLAB is a popular toolbox used for image processing. However it can be done using other software and codes can be written using various programming languages. Python is one of the languages which can be used for image processing.

2.1 Introduction

Image processing is a widely explored research field, as it finds applications in various domains. Some of the most common application areas are remote sensing, astronomy, military, defense, nuclear medicine, industry, surveillance, law enforcement, medical science, weather forecasting and many more [1, 2]. For all the applications, images are required that contain important information. This information may be in terms of fine structures, edges, ridges, contours, colours and other details [3]. Sometimes it is difficult to retrieve all the information from images so there is a need to process the images in such a manner so that details of images can be retained while unwanted signals are suppressed or removed from them. For this purpose some manipulation is to be done by applying various filters and operators. This manipulation is called 'image processing'. Prior to delving into the concept of image processing, it is essential to comprehend the definition of an image [4]. In our day-to-day life, we all take pictures using a camera. In the early days cameras were analog, but in the present day digital cameras are to be used. Any analog signal can be changed into a digital signal using sampling and quantization techniques. A digital image represents an image as a set of numbers, which is handled and stored in a digital computer. A digital image is divided into small elements called picture elements or pixels. Pixels are the smallest part of any digital image and they are arranged in matrix form. Each pixel has two coordinates, x and y, and is represented as $f(x, y)$ where f is the integer value from 0 to 255 for a gray image. 0 represents black while 255 represents white pixels. Between 0 to 255 gray pixels exist with different intensities [4].

Image processing is a method in which an image is manipulated in such a way that its quality is improved according to the desire of the applications. For example, in medical imaging systems like ultrasound images [3], computed tomography image [2], magnetic resonance image; noise introduces during the acquisition of the medical images due to various phenomena [4]. Ultrasound images can be affected by speckle noise, which serves as an illustration of multiplicative noise [1, 2]. It is multiplied with random pixels by different amounts. Speckle is the inherent property of ultrasound images that is introduced due to the diffraction of ultrasound signals used by the imaging system. A computed tomography image may be distorted by Gaussian noise, which falls into the category of additive noise. It is added with the pixel values and corrupts the image. MRI images are affected by Rician noise, whose characteristic is entirely different from speckle and Gaussian noise [1–3]. Based on the preceding discussion, it is evident that noise signals can impact images, leading to a degradation in image quality and the suppression of valuable information. Suppression of noise signals is one of the most important preprocessing tasks and is called image denoising. Denoising is done by applying various operators on to the images according to the character tics of noise signals. This technique is called filtration, which is done using filters [5]. However, image denoising is not only the manipulations; there are other different techniques which can be used to extract important information. Image processing is commonly categorized into three levels; low-level, medium-level, and high-level image processing [4].

Low level image processing: In this level both the inputs and outputs are images. The following processes come under this category:

- Image denoising
- Edge detection
- Contrast enhancement

- Sharpening of images

Medium level image processing: In this level inputs are images but outputs are image attributes. For example:
- Image segmentation
- Object recognition
- Description of images

High level image processing: This level involves making sense from a group of recognized objects. It is associated with computer vision.

Based on the preceding discussion, image processing is very vast stream and it is widely used in various application areas [6]. Hence it is not possible to explain all the topics in a single chapter. In this chapter image, processing operations using Python language have been discussed. Image processing and computer vision rely on a set of algorithms, and the main package of skimage offers a limited number of utilities. For more features; one of the following sub-packages (Data, Color, Exposure, Filters, etc.) must be imported. https://scikit-image.org/docs/dev/api/skimage.html is the library used to import the images and data from it.

2.2 Python and its libraries for image processing

Python was created by Guido Van Rossum and is a high-level, general-purpose programming language. It can be used to write programming codes for small and large scale projects. It is easy to understand for beginners [7]. It supports both object-oriented and functional types of programming. The Python Software Foundation is a global community that manages and directs resources for development of python. Python has following characteristics:
- It is a general purpose programming language.
- It comes under the category of high-level programming.
- It has a comprehensive standard library.
- In this program codes can be written for both small- and large-scale projects.
- It supports object-oriented programming.
- It also supports functional programming.

There are many libraries in Python; among these some of the libraries may be used to process image data and analyze it. OpenCV is one of them.

2.2.1 OpenCV

OpenCV, an abbreviation for Open Source Computer Vision Library, is a freely accessible software library that primarily focuses on OpenCV, short for Open Source Computer Vision Library, serves as a freely accessible software library with a primary emphasis on computer vision and machine learning tasks. Its main objective revolves around delivering a comprehensive framework tailored for computer vision applications and facilitating the seamless integration of machine perception into commercial products. OpenCV is distributed under BSD license, allowing businesses to conveniently employ and modify its source code according to the specific require-ments. The library provides an extensive range of optimized algorithms, exceeding 2500 in number. These algorithms encompass both classical and state-of-the-art techniques, making the library a comprehensive resource for computer vision and machine learning tasks. These algorithms empower users with a wide array of capabilities, including but not limited to face detection, recognition, object identi-fication, human action classification in videos, camera motion tracking, object tracking, extraction of 3D object models, generation of 3D point clouds from stereo cameras, creation of high-resolution panoramic images via image stitching, database image search, red-eye removal in flash photography, eye movement tracking, scene recognition, marker placement for augmented reality applications, and more. OpenCV has a large and active user community, with over 47 000 individuals and an estimated download count exceeding 18 million. It is widely adopted by companies, research groups, and governmental organizations. Link to OpenCV website https://opencv.org/about/. This library offers an extensive range of features that surpass expectations. It provides numerous capabilities, including image display and restora-tion after applying various filters. Its power and versatility are truly remarkable.

2.2.2 NumPy

NumPy holds a vital position as a foundational package in Python's scientific computing ecosystem. It encompasses a wide spectrum of functionalities, including a robust N-dimensional array structure, advanced broadcasting functions, and seam-less integration utilities for incorporating C/C++ and Fortran code. Additionally, NumPy offers essential capabilities for tasks like linear algebra, Fourier transform operations, and random number generation. Although its main focus is on scientific applications, NumPy can efficiently handle multi-dimensional data for various purposes. It supports the definition of arbitrary data types, enabling seamless integration with diverse databases and promoting efficient data processing. Its URL is https://www.numpy.org/. We utilize this library to apply filters to images, treating them as matrices. As stated previously, an image can be represented as a multidimensional matrix. Python lists different functions as arrays capable of storing various types of elements [8]. In Python, an object can be thought of as a reference to a memory location that contains all the object's attributes, including its data and values. While this additional information enables Python's dynamic typing, it incurs a cost, especially when dealing with large collections of objects like arrays. The structure of Python lists involves an array of pointers where each pointer refers to a specific location that holds information about an element [9]. Python lists have some limitations in terms of memory and computation overhead. When elements stored in a list share the same type, there is significant redundancy in the information stored. To overcome this, NumPy arrays are utilized. NumPy arrays exclusively store homogeneous elements, meaning elements of the same data type. This attribute renders objects more efficiently in terms of storage and manipulation than Python lists, particularly when handling a substantial volume of elements.

NumPy arrays excel in scenarios where the array contains a substantial number of elements, such as thousands or millions [10]. They offer efficient storage and computational capabilities. Moreover, NumPy arrays allow for element-wise operations, which are not possible with Python lists. This advantage makes them a preferred choice when performing mathematical operations on extensive datasets. Generating NumPy arrays is a straightforward process, despite the complex problems they can solve. To create a simple ndarray, you can utilize np.array () function. Just pass the array values as a list when invoking the method:

```
import numpy as np

#It generates one dimension array.

array1 = np.array([1, 4, 6, 8])
print("output")
print(array1)
```

After running this code the output is

```
[6 1 4 8]
```

```
# 1-D array with specified data type

array2 =np.array([1, 4, 6, 8], dtype=np.float32)
print("output")
print(array2)
```

After running this code the output is

```
[6. 1. 4. 8.]
```

```
# It generate the multi-dimensional array.

array3 =np.array([1, 4, 6, 8],[1, 2, 7, 9]])
print("output")
print(array3)
```

After running this code the output is

```
[[6 1 4 8]
 [7 2 9 1]]
```

2.3 Various operations for image processing using different libraries

Image processing requires various types of operations, e.g., loading, reading, writing, displaying, saving the images, etc. These operations can be performed for

Figure 2.1. Sample image.

grayscale as well as color images. Some of the other important operations are cropping, resizing, translation, and rotation. Below, all the aforementioned operations are discussed one by one.

- **Loading of images using skimage**

The data module of the scikit-image package includes sample images. Let us consider that single image is to be loaded to perform some experiments. This image may be loaded from the package. Python code to perform the above task is given below:

```
#Load Image from skimage

from skimage . io import imread, imshow
from skimage import data
image = data.flowers()

imshow(image)
```

The output (figure 2.1) is an image as shown below:
The imshow function is utilized to view the image.

- **Reading images using skimage**

If you would rather upload an image from your device instead of using pre-existing ones in the package, you can make use of the imread function provided by skimage. This function enable us to read images in two formats: colored and grayscale. By examining both formats, we can enhance our comprehension of their differences. To accomplish this, the imread function includes a parameter named 'as-gray', which can be used to specify whether the image should be converted to grayscale. Let's start by reading an image in grayscale format by setting the `as_gray` parameter to true:

```
#Load Image from Desktop

import skimage.io as io
import matplotlib.pyplot as plt

image_path = 'flowers.jpg'
gray_image = io.imread(image_path, as_gray=True)

plt.imshow(gray_image, cmap='gray')
plt.axis('off') # Optional: Turn off axis labels
plt.show()
```

The output (figure 2.2) is an image as shown below.

Our next step involves loading the image in its original color format. For achieving this, we must set the `as_gray` parameter to False.

```
#Load Image from Desktop

from skimage.io import imread, imshow
import matplotlib.pyplot as plt

image_path = 'flowers.jpg'
color_image = imread(image_path, as_gray=False)

print(color_image.shape)
```

Figure 2.2. Gray image.

Figure 2.3. Color image.

```
imshow(color_image)
plt.show()
```

The output (figure 2.3) of the above code is shown below.
- **Reading, writing and displaying images using Numpy**

Machines rely on numerical representations to perceive and process various forms of data, including images and text. You might be wondering how images are converted into numerical values. The answer lies in pixel values, where each number corresponds to the intensity of a pixel at a specific location. In the provided grayscale image, the pixel values indicate the intensity of black at each corresponding position. Color images consist of multiple values per pixel, representing the intensity of the respective Red, Green, and Blue channels in the case of RGB images. The ability to read and write images is essential for any computer vision project, and OpenCV greatly simplifies this task. Now, let's delve into the process of importing an image into our machine using OpenCV.

```
#import the libraries
import numpy as np
import matplotlib.pyplot as plt
import cv2

# Reading the image
image = cv2.imread('flowers.png')
image = cv2.cvtColor(image, cv2.COLOR_BGR2RGB)

# Writing the image
cv2.imwrite('flowers.jpg', cv2.cvtColor(image, cv2.
```

```
COLOR_RGB2BGR))

plt.show()
```

In the above codes BGR stands for Blue, Green and Red. In this format the Blue channel comes first, the Green channel comes second and the Red channel comes third, while RGB stands Red, Green and Blue. In this format Red comes first, Green comes second and Blue comes third. Function BGR2RGB and RGB2BGR are used to transform from one format to another.

- **Changing color spaces using Numpy**

A color space is a standardized system that represents colors in a way that facilitates accurate reproduction. In the context of images, grayscale images consist of single pixel values, while color images consist of three values per pixel, representing the intensities of the Red, Green, and Blue channels (RGB). While RGB is the primary color space for many computer vision applications, certain use cases such as video compression and device-independent storage rely on alternative color spaces like Hue-Saturation-Value (HSV). OpenCV, by default, reads images in the BGR format. Therefore, when working with OpenCV, it is necessary to convert the color space of the image from BGR to RGB [4]. Let's explore how to perform this conversion:

```
#import the required libraries

import numpy as np
import matplotlib.pyplot as plt
import cv2

# Read the image
image = cv2.imread('flowers.jpg')

# Convert image to grayscale
gray_image = cv2.cvtColor(image, cv2.COLOR_BGR2GRAY)

# Plot the grayscale image
plt.subplot(121)
plt.imshow(gray_image, cmap='gray')
plt.title('Grayscale Image')

# Convert image to HSV format
hsv_image = cv2.cvtColor(image, cv2.COLOR_BGR2HSV)
```

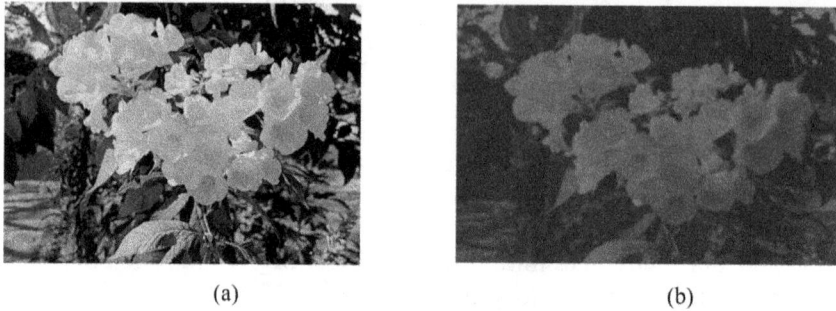

(a)   (b)

Figure 2.4. Same image in different space (a) grayscale image, (b) HSV image.

```
# Plot the HSV image
plt.subplot(122)
plt.imshow(hsv_image)
plt.title('HSV Image')
plt.show()
```

The output (figure 2.4) of the above code is shown below.

2.3.1 Read, display and save operations using OpenCV

• **Reading image**

To utilize the OpenCV library, we first need to install it in our 'miasi' environment. To do this, execute the following command: 'pip install opencv-python'. Once the installation is complete, we can proceed with our tasks. In OpenCV, images can be read by various methods. However, we will primarily concentrate on three main methods:

```
import cv2

image = cv2.imread("flowers.jpg", cv2.imread_color)
```

This method is used to load a color image. It ignores any transparency present in the image. It is the default flag if no other flag is specified.

```
import cv2

image = cv2.imread("flowers.jpg", cv2.imread_grayscale)
```

This method is used to load the image in grayscale mode, converting it to black and white.

```
import cv2

image = cv2.imread("flowers.jpg", cv2.imread_unchaged)
```

This method loads the image as it is, including the alpha channel if it exists. These methods offer different approaches to read and interpret images in OpenCV, accommodating diverse requirements and use cases.

```
import cv2

#reading images from directory file
img1= cv2.imread('flowers.jpg', -1)
print('img1 Shape :{}'.format(flowers.shape))
print("Height: {} pixels" .format(flowers.shape[0]))
print("width:{} pixels".format(flowers.shape[1]))
print("channels: {}".format(stones.shape[2]))
print("size", flowers.size)
print("Type: ", type(flowers))
Image1 Shape (350, 512, 3)
Height: 350 pixels
Width: 512 pixels
channels: 3
size: 537600
```

- **Reading Image with OpenCV**

The `imread()` function in OpenCV enables the conversion of an image into a NumPy array.
  - **Display an image using OpenCV**

To display an image in a window using OpenCV, you can use the `cv2.imshow()` function. This function takes two arguments: the window name as a string and the image itself. It automatically adjusts the window size to fit the image.

To enable keyboard binding in OpenCV, you can use the `cv2.waitKey()` function. This function allows you to specify the duration in milliseconds for rendering the image. It waits for the specified duration or until a keyboard event occurs. The program pauses at this point, waiting for a key press to continue. If no argument or 0 is passed to `cv2.waitKey()`, it will wait indefinitely until a keyboard event is detected.

To close the windows created by you `cv2.destroyAllWindows()` function can be used which destroys all the windows at once but if you want to close a particular window, you can use `cv2.destroyWindow()` function. Window name must be provided as argument.

```
cv2.imshow('Original image', flowers)
cv2.imshow('Grayscale image', stones)
cv2.waitKey()
cv2.destroyAllWindows ()
```

- **Saving image using OpenCV**

To save an image using OpenCV, you can utilize the `cv2.imwrite()` function. This function takes two arguments: the desired file name where you want to save the image and the image data itself. For example, you can save an image named `output.png` by using the following code: `cv2.imwrite ('output.png', img1)`. Make sure to use the appropriate file name and extension based on your requirements. This code will save the image in the PNG format within the current working directory.

- **Color image space using OpenCV**

Color space is a fundamental concept in computer vision and image processing, as it defines a specific method of organizing and representing colors. A color space comprises a color model and a mapping function, which collaborate to determine how pixel values are represented using tuples. Different color spaces serve different purposes and offer unique characteristics. Some frequently utilized color spaces include RGB, HSV, YUV, and Lab.

RGB (Red, Green, Blue) is a widely recognized and popular color space. It represents colors by combining red, green, and blue channels. In the RGB color model, each pixel value is represented as a tuple of three numbers, typically ranging from 0 to 255, which indicate the intensity of each color channel.

HSV (Hue, Saturation, Value) is a cylindrical color space that separates the fundamental characteristics of colors: hue, saturation, and value. This color space closely aligns with the human visual system's perception of colors. HSV provides flexibility in image processing and analysis, allowing for intuitive adjustments of hue, saturation, and brightness.

It's worth noting that there is a distinction in pixel ordering between OpenCV and Matplotlib. OpenCV follows the BGR (Blue, Green, Red) order for color representation, while Matplotlib typically follows the RGB (Red, Green, Blue) order.

```
px = img1[40, 75]
print (px)
(B, G, R)=img1[40, 75]
print('R = {}, G = {}, B = {}'.format(R,G,B))
```

The output will of the above code is:

```
[165 170 175]
R = 165, G = 170, B = 175
```

- **Changing color space using OpenCV**

In OpenCV, the color conversion process involves using the `cv2.cvtColor()` function. This function takes the input image and a flag as parameters to specify the desired type of conversion. To convert an image from BGR to grayscale, you can use the flag `cv2.COLOR_BGR2GRAY`. On the other hand, to convert an image from BGR to RGB, you can use the flag `cv2.color_bgr2rgb`. Similarly, if you want to convert an image from BGR to HSV, you can use the flag `cv2.color_bgr2hsv`.

```
img3 = cv2.cvtColor(flowers, cv2.color_bgr2rgb)
img4 = cv2.cvtColor(flowers, cv2.color_bgr2hsv)
cv2.imshow("Image hsv", img3)
cv2.imshow("Image rgb", img4)
cv2.waitKey(0)
```

The output (figure 2.5) of the above program is shown below.

- **Cropping image using OpenCV**

To crop an image using OpenCV, you can utilize array slicing. By specifying the desired startY and endY coordinates, as well as the startX and endX coordinates, you can extract the specific region of interest from the image. This process effectively crops the image, allowing you to focus on the selected portion.

```
crop = flower[ 20 : 400, 25 : 100 ]
cv2.imshow("cropped image", crop)
cv2.waitKey(0)
```

(a)   (b)

Figure 2.5. Same image in different space (a) Image HSV, (b) Image RGB.

Figure 2.6. Cropped image.

The output (figure 2.6) is shown below.

- **Resizing the image using OpenCV**

Resizing an image refers to altering its dimensions, whether it's the width, height, or both, while preserving the aspect ratio if desired. OpenCV provides the cv2.resize() function for image resizing. When shrinking an image, it is recommended to use the interpolation method cv2.inter_area. On the other hand, for zooming, cv2.inter_cubic (slower) or cv2.inter_linear can be used. The default interpolation method for resizing in OpenCV is cv2.inter_linear.

```
height, width= img1.shape[ :2]
res1 = cv2.resize(img1, (width+ 100, height - 100),
interpolation = cv2.inter_cubic)
res2 = cv2.resize(img1, (width // 2, height // 2),
interpolation = cv2.inter area)
res3 = cv2.resize(img1, (width*2, height))
```

```
cv2.imshow("Scaling inter cubic", res1)
cv2.waitKey(0)
cv2.imshow("Scaling inter area", res2)
cv2.waitKey(0)
cv2.imshow("Scaling inter linear", res3)
cv2.waitKey(0)
```

- **Translation using OpenCV**

Translation in image processing refers to shifting the image by adjusting the x and y coordinates. This can be accomplished using a transformation matrix, typically denoted as **M**:

```
M = [ [ 1, 0, tx ],
[0, 1, ty ] ]
```

In the transformation matrix, `tx` and `ty` represent the translation values for the x and y-directions, respectively, determining the amount of shift applied horizontally (rightward) and vertically (downward) to the image. Once the transformation matrix is constructed, we can use the `cv2.warpAffine()` function to apply the translation to our image. The third parameter in `cv2.warpAffine()` defines the size of the output image after the transformation.

```
import cv2
import numpy as np

M = np.float32([[ 1, 0, 100 ], [ 0, 1, 50 ]])
dst = cv2.warpAffine (img1, M, (width, height))
cv2.imshow("translate image", dst)
cv2.waitKey(0)
```

Rotation of image using OpenCV

To perform image rotation in OpenCV using Python, you can follow these steps. First, calculate the affine matrix that represents the desired transformation, including the rotation. Then, apply the affine matrix to the input image using the `warpAffine()` function. The `cv2.get Rotation Matrix2D()` function can be used to create the affine matrix. It takes the center point around which the image will be rotated as the first argument, followed by the rotation angle in degrees, and an optional scaling factor. Rotation is a type of transformation that can be achieved by using the following transformation matrix:

$$R = \begin{bmatrix} \cos\theta & -\sin\theta \\ \sin\theta & \cos\theta \end{bmatrix}$$

where θ represents the angle of rotation in the counter clockwise direction.

Figure 2.7. Rotation.

```
Import cv2

height, width = img1.shape[ :2]
R = cv2.getRotationMatrix2D(( width/2, height/2), 30,
0.8)
dst = cv2.warpAffine (img1, R, (width, height))
cv2.imshow('rotation', dst)
cv2.waitKey(0)
```

The output (figure 2.7) image is shown below.

2.3.2 Applications of image processing in OpenCV

In this section, we will explore the application of visual effects to images. We will cover the usage of essential image processing operators, such as edge detection, histogram computation, template matching, and the application of image filters. These techniques enable us to create various effects in photographs.

- **2D convolution**

Convolution plays a crucial role in signal processing and image processing. It involves applying a kernel or filter to a signal or image to achieve different effects. In the digital domain, convolution is performed by multiplying and accumulating overlapping samples of two input signals or matrices. In the context of image processing, a 2D matrix representing an image is convolved with a smaller matrix known as a kernel or image filter. The process of applying the kernel to an image is called image filtering, and it results in a filtered image. Different types of filters can be used, such as low-pass filters (LPFs) for noise reduction or blurring, and high-pass filters (HPF) for edge detection. OpenCV provides the `cv2.filter2D()` function to perform convolution between a kernel and an image. For example, we can apply an averaging filter to an image using a 3×3 kernel.

Figure 2.8. Result of 2D convolution (a) original image (b) filtered image.

$$K = \frac{1}{9} \begin{bmatrix} 1 & 1 & 1 \\ 1 & 1 & 1 \\ 1 & 1 & 1 \end{bmatrix} \qquad (2.1)$$

```
import cv2
import numpy as np

img = cv2.imread("flowers.png")
kernel = np.ones((3, 3), np.float32)/9
dst = cv2.filter2D (img, -1, kernel)

cv2.imshow("Original image", img)
cv2.imshow("Filtered image", dst)
cv2.waitKey(0)
cv2.destroyAllWindows()
```

The output (figure 2.8) images are shown below.
- **Smoothing image**

Image blurring is a technique used to reduce noise in an image by convolving it with a low-pass filter kernel. This process effectively eliminates high-frequency components, including noise and edges. As a result, there is a blurred appearance of the edges when the filter is applied.
- **Averaging**

OpenCV provides the cv2.filter2D() function for applying convolution between an image and a kernel. To illustrate, we will apply an averaging filter to an image. Let's define a 5×5 averaging filter kernel as shown below:

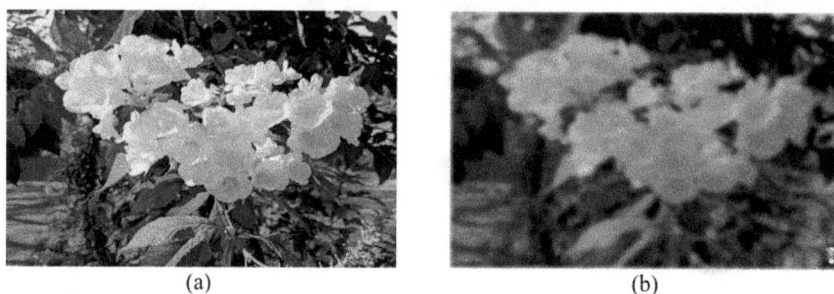

(a)   (b)

Figure 2.9. Result of averaging (a) original image (b) blurred image.

$$K = \frac{1}{25} \begin{bmatrix} 1\,1\,1 & 1\,1 \\ 1\,1\,1 & 1\,1 \\ 1\,1\,1 & 1\,1 \\ 1\,1\,1 & 1\,1 \\ 1\,1\,1 & 1\,1 \end{bmatrix} \tag{2.2}$$

```
import cv2

Import numpy as np

blur= cv2.blur( flowers, (5, 5))
cv2.imshow("Original image", img)
cv2.imshow("Blurred image", blur)
cv2.waitKey(0)
cv2.destroyAllWindows ()
```

The output (figure 2.9) images are shown below.
- **Gaussian filtering**

Images can contain different types of noise, often originating from the source such as a camera sensor. To reduce this noise, image smoothing techniques are employed. OpenCV offers several methods for image smoothing, also known as blurring. One effective approach is using Gaussian filters, which possess the advantage of eliminating overshoot in response to a step function input while minimizing abrupt transitions. Gaussian smoothing smoothens sharp edges in images without excessively blurring them. To apply Gaussian smoothing to an input image, OpenCV provides the cv2. gaussianblur() function.

(a)   (b)

Figure 2.10. Result of Gaussian filtering (a) original image (b) Gaussian blurred image.

```
blur= cv2.GaussianBlur (img1, (3, 3), 0)
cv2.imshow("Original image", img)
cv2.imshow("Gaussian blurred image", blur)
cv2.waitKey(0)
cv2.destroyAllWindows()
```

The output (figure 2.10) images are shown below.
- **Median filtering**

The cv2.medianBlur() function is employed to compute the median value of all the pixels within the kernel window. This median value is subsequently utilized to replace the central pixel. This approach proves especially effective in removing salt-and-pepper noise. It's worth noting that, unlike Gaussian and box filters that can produce filtered values not present in the original image, median filtering always substitutes the central element with a pixel value from the original image. This property contributes to its noise reduction capability. To ensure proper functioning, it is crucial to use a kernel size that is a positive odd integer.

```
import cv2

median = cv2.medianblur (flowers, 4)
cv2.imshow("Original image", flowers)
cv2.imshow("Median filtering image", median)
cv2.waitKey(0)
cv2.destroyAllWindows()
```

- **Sharpening**

The degree of sharpening achieved is determined by the choice of kernel. Customization of the kernel grants significant flexibility and allows for various types of sharpening effects. For general sharpening purposes, a kernel like the following can be used:

$$M = \begin{bmatrix} -1 & -1 & -1 \\ -1 & 5 & -1 \\ -1 & -1 & -1 \end{bmatrix} \qquad (2.3)$$

For a more pronounced sharpening effect, an alternative kernel that can be utilized is as follows:

$$M = \begin{bmatrix} 1 & 1 & 1 \\ 1 & -5 & 1 \\ 1 & 1 & 1 \end{bmatrix} \qquad (2.4)$$

```
import cv2
import numpy as np

cv2.imshow("Original image", img)

# generating the kernels
kernel_sharpen1 = np.array([[-1, -1, -1], [-1, 5, -1],
[-1, -1, -1]])
kernel_sharpen2 = np.array([[1, 1, 1], [1, -5, 1], [1, 1,
1]])

# applying different kernels to the input image
output1 = cv2. filter2D (img, -1, kernel_sharpen1)
output2 = cv2.filter2D (img, -1, kernel_sharpen2)

cv2.imshow("Sharpening", output1)
cv2.imshow("Excessive Sharpening", output2)
cv2.waitKey(0)
```

- **Edge detection**

Edge detection is a commonly employed method in image processing, focused on recognizing the boundaries or edges of objects within an image. It operates by detecting abrupt changes or discontinuities in brightness values. Edge detection is of utmost importance in fields like computer vision, image processing, and machine vision, particularly in tasks related to image segmentation and data extraction. There are various algorithms that can be used for edge detection, such as Sobel, Canny, Prewitt, Roberts, and fuzzy logic methods. The Laplacian operator is one such algorithm that highlights regions of rapid intensity changes, making it suitable for edge detection. Mathematically, the Laplacian of an image $I(x,y)$, representing its intensities, can be calculated using the following formula:

(a)   (b)

Figure 2.11. Result of edge detection (a) original image (b) Laplacian.

$$L(x, y) = \frac{\partial^2 f}{\partial x^2} + \frac{\partial^2 f}{\partial y^2} \tag{2.5}$$

In the discrete approximation of the Laplacian operator for a specific pixel, the calculation entails determining the weighted average of the intensities of its neighboring pixels. This operation aims to capture the intensity variations and detect edges within the image.

$$K = \begin{bmatrix} 0 & 1 & 0 \\ 1 & -4 & 1 \\ 0 & 1 & 0 \end{bmatrix} \tag{2.6}$$

```
import cv2

image = cv2.imread("flowers.jpg")
laplacian = cv2.Laplacian (image, cv2.CV_64F)
cv2.imshow('Original image', image)
cv2.imshow('Laplacian, laplacian)
cv2.waitKey(0)
```

The output (figure 2.11) images are shown below.
- **Canny edge detection**

The Canny edge detection algorithm, first proposed by John F Canny in 1986, is a popular technique for detecting edges in images. It follows a multi-stage approach consisting of several essential steps, which include:
- Noise suppression
- Calculation of gradient
- Non-maximum suppression
- Double thresholding
- Edge tracking

(a)   (b)

Figure 2.12. Result of Canny edge detection (a) original image (b) Canny edge detection.

It's important to mention that the Canny algorithm works with grayscale images, so it is necessary to convert the image to grayscale before applying the aforementioned steps.

```
import cv2

image = cv2.imread ('flowers.jpg')
canny = cv2.Canny (image, 40, 250)

cv2.imshow('Original image', img)
cv2.imshow('Canny', canny)
cv2.waitKey(0)
cv2.destroyAllWindows()
```

The output (figure 2.12) images are shown below.

2.4 Conclusion

Based on the above discussion, it is evident that Python is a programming language suitable for writing image processing code. Python offers simplicity and beginner-friendly syntax, making it easy to understand and write code for image processing tasks.

References

[1] Gonzalez R C and Woods R E 2009 *Digital Image Processing* 3rd edn (Englewood Cliffs, NJ: Prentice-Hall)
[2] Bhonsle D, Chandra V K and Sinha G R 2018 *An optimized framework using adaptive wavelet thresholding and total variation technique for de-noising medical images J. Adv. Res. Dyn. Control Syst.* **10** 953–65
[3] Bhonsle D, Chandra V K and Sinha G R 2018 *Speckle noise removal from ultrasound images using combined bivariate shrinkage and enhanced total variation techniques Int. J. Pure Appl. Math* **118** 1109–31

[4] Bhonsle D, Chandra V K and Sinha G R 2017 *Noise removal from medical images using shrinkage based enhanced total variation technique J. Adv. Res. Dyn. Control Syst.* **13** 549–60

[5] Bhonsle D 2021 Quality improvement of Richardson lucy based de-blurred images using krill herd optimization *IEEE Int. Conf. on Advances in Electrical Computing Communication and Sustainable Technologies* pp 1–5

[6] Bhonsle D, Chandra V K and Sinha G R 2012 Medical image denoising using bilateral filter *Int. J. Image Graphic. Signal Process.* **4** 36–43

[7] Zamfir F S and Pricop E 2022 *On the design of an interactive automatic Python programming skills assessment system 14th Int. Conf. on Electronics, Computers and Artificial Intelligence (ECAI) (Ploiesti, Romania, 2022)* pp 1–5

[8] Python programming language 2020 https://python.org

[9] Jaworski M 2021 *Tarek Ziadé, Expert Python Programming: Master Python by Learning the Best Coding Practices and Advanced Programming Concepts* (Birmingham: Packt Publishing)

[10] Lorandi Medina A P, Ortigoza Capetillo G M, Saba G H, Perez M A H and Garcia Ramirez P J 2020 *A simple way to bring Python to the classrooms IEEE Int. Conf. on Engineering Veracruz ICEV 2020*

IOP Publishing

Image Processing with Python
A practical approach
Irshad Ahmad Ansari and Varun Bajaj

Chapter 3

Review and implementation of image segmentation techniques in Python

Siddharth Bhalerao, Sourabh Sahu and Saurabh Tewari

Image segmentation is one of the central techniques in image processing. It is important as it forms basis of various other computer vision and image processing algorithms. Segmentation finds its uses in various fields like automation, medical image processing, satellite image processing. Segmentation refers to partitioning of an image in order to extract useful information from it. There are several techniques available for image segmentation. There is no single technique that could cater to all segmentation tasks. All techniques have their pros and cons. This chapter deals with three different image segmentation techniques. The thresholding-based image segmentation is introduced first followed by discussion of edge detection algorithms in images. Machine-learning based image segmentation techniques are introduced with a comprehensive discussion on K-Means clustering-based image segmentation. A step-by-step process is provided for the practical implementation of all the techniques in Python.

3.1 Introduction

Image segmentation is a technique used in computer vision that partitions an image into several discrete segments or regions, allowing for an enhanced and in-depth analysis of the image [1]. The primary objective of this process is to transform the image representation into a more meaningful and analyzable form. It enables us to locate objects within the image and establish clear boundaries between them [2].

It is not always efficient or necessary to process an entire image as a single unit, as certain regions of the image may not contain relevant or useful information. Consequently, image segmentation enables us to focus selectively on the crucial parts pertinent to our analysis or task, improving efficiency and effectiveness. Image segmentation has diverse applications across various industries. At its core, an image

doi:10.1088/978-0-7503-5924-5ch3

is composed of individual pixels. In image segmentation, pixels with similar attributes, such as color or intensity, are grouped.

Image segmentation has a large number of applications in different domains. Image segmentation is highly valuable in the healthcare domain [3]. It aids in segmenting cancer cells and tumors, allowing for accurate diagnosis, treatment planning, and monitoring of the severity of the disease [4, 5]. Image segmentation also finds its applications in fields like robotics, video surveillance, remote sensing [6, 7], vegetation detection [8, 9], etc. It provides crucial insights by extracting meaningful information from images. Image segmentation plays a critical role in enabling autonomous driving. Object detection, which involves segmentation, is essential for identifying and understanding the surrounding objects and obstacles on the road. Semantic image segmentation is another very popular research domain [10]. Semantic image segmentation is used to identify different objects in an image by providing them labels [11, 12]. Similarly, superpixel segmentation is one of the growing areas of research [13]. In superpixel segmentation, a region is identified in the image that has a large number of pixels with a meaningful relationship. The relationship could be based on color, texture, etc. The pixel grouping is termed a superpixel [14–16].

In this chapter, different categories of image segmentation techniques and their practical implementation have been discussed. In the next section, threshold-based image segmentation is described and a popular segmentation technique, Otsu's segmentation algorithm, is covered [17]. Line, point, and edge detection is one of the core areas of image segmentation. In section 3.3, discussions on edge detection techniques are present. Machine learning-based segmentation techniques are discussed in section 3.4.

3.2 Thresholding based image segmentation

Thresholding-based segmentation is one of the most popular and simplistic methods for image segmentation [18]. Considering a grayscale image, a thresholding algorithm calculates a threshold which is nothing but a brightness level that divides an image into two intensity levels. Therefore, the result of such thresholding is a binary image and all the gray levels lower than or equal to the threshold are made 0 and all levels greater than the threshold will be made 1. Intuitively it can be understood that the image is getting divided into two parts: foreground and background [2]. The idea behind thresholding is explained in equation (3.1).

$$I'(x, y) = \begin{cases} 1, & if \quad I(x, y) > T \\ 0, & if \quad I(x, y) \leqslant T \end{cases} \tag{3.1}$$

In the above equation $I(x, y)$ is the grayscale image, x, y denotes the coordinates of a pixel and T is the threshold. The resultant image $I'(x, y)$ is the binary image with two regions. The thresholding scheme could be extended to multiple threshold scenario and in such case the technique is termed as multilevel thresholding. Multilevel thresholding divides an image into multiple regions [19]. Thus, a multi-level thresholding algorithm providing four thresholds will divide an image into five regions. When an entire image is thresholded using a single threshold value, such thresholding is called global thresholding. Similarly, when the threshold is different for different regions of the image such thresholding is called as local thresholding. Next sub-section discusses Python implementation of Otsu's method which is one of the most popular thresholding methods.

3.3 Python implementation

In this sub-section a brief description of Otsu's thresholding scheme is provided. The Otsu's technique utilizes image histogram for calculating the optimal threshold for a grayscale image. Histogram of a digital image with L brightness levels, in range $[0, 1, 2, \ldots, L - 1]$ is a function $h(i) = n_i$ where i is the brightness level and n_i is the number of pixels having i as brightness level. The probability distribution is computed from image histogram using equation (3.2).

$$p_i = \frac{h(i)}{N} \tag{3.2}$$

The p_i is the probability histogram and N is the total number of pixels in the image, if image resolution is $m \times n$ then $N = m \times n$. The threshold $T(k)$ is the value of brightness level that divides a whole image in two classes. The class probability for a class is expressed by the following equation:

$$\omega(k) = \sum_{i=0}^{k} p_i \tag{3.3}$$

Similarly, class mean is calculated by equation (3.4).

$$\mu(k) = \sum_{i=0}^{k} i p_i \tag{3.4}$$

In the above equations, k could be any value between $[0, L - 1]$. The optimal threshold value maximizes the between class variance calculated by following equation:

$$\sigma_B^2(k) = \frac{[\mu_T \omega(k) - \mu(k)]^2}{\omega(k)[1 - \omega(k)]} \tag{3.5}$$

In the above equation, μ_T is the total probability mean of the image, and is calculated as following:

$$\mu_T = \sum_{i=0}^{L-1} i p_i \tag{3.6}$$

The optimal threshold $k*$ is that value of k which maximizes between class variance calculated in equation (3.5). Hence,

$$\sigma_B^2(k*) = \max_{0 \leqslant k < L-1} \sigma_B^2(k) \qquad (3.7)$$

The value obtained in equation (3.7) is the optimal threshold $k*=T$ and using (3.1) the image is segmented into two regions. The Python program of Otsu's method is implemented and explained in the following code snippets.

```python
import cv2
import numpy as np
import matplotlib.pyplot as plt
```

The Otsu's scheme is implemented using the OpenCV library on Python. OpenCV offers various image processing functionalities. The implementation uses NumPy library for array operations. The well-known library Matplotlib is used for image visualization.

```python
# read image file 3.jpg
img = cv2.imread('images\\3.jpg')

# convert color image to grayscale
img = cv2.cvtColor(img,cv2.COLOR_BGR2GRAY)

# resize image 256x256
img = cv2.resize(img,(256,256))

[m,n] = img.shape

#calculate histogram
h = cv2.calcHist([img],[0],None,[256],[0,256])

# output histogram is list of lists
# using hstack to convert to np array
h = np.hstack(h)

# plot image and its histogram
plt.figure(figsize=(10,5))
plt.subplot(1,2,1)
plt.imshow(img, cmap = 'gray')
plt.axis('off')
plt.subplot(1,2,2)
plt.plot(h)
plt.show()
```

(a)   (b)

Figure 3.1. (a) Grayscale image, (b) image histogram.

In the above snippet, the imread() function from OpenCV library is used for reading the image. The Otsu's scheme discussed earlier works on grayscale images, therefore, function cv2.cvtColor() is used to convert it to grayscale and the image is then resized to 256 × 256 resolution. The most important part of Otsu's technique is the histogram. The cv2.calcHist() function from the OpenCV library provides a histogram of the image. It requires five arguments for computation of the histogram. The first argument is the image itself passed on as a Python list, the second argument is the channel. This argument is 0 as the grayscale image contains only one channel. The third argument is a mask; it is *None* as all pixels must be included in computation of the histogram, therefore, no masking is required. The fourth argument is the number of bins required in the histogram; as the brightness levels in gray images include [0, 255], this argument is 256. The last argument indicates the range of the histogram, which is starting from 0 and going up to 255.

After computation of the histogram, the Matplotlib library is used for visualizations of the image and its histogram. The subplot function is used to plot or draw multiple plots/images into a single plot. The output of the above Python snippet is present in figure 3.1(a) and its histogram is shown in figure 3.1(b). The histogram has 256 bins with pixel values starting from 0 and going up to 255.

```
p = h/np.sum(h)

L = 255

#intializing the variables
omega = np.zeros(L+1)
mu = np.zeros(L+1)
Sum = 0

# computing omega and mu for complete image
for i in range(L+1):
    omega[i] = np.sum(p[0:i+1])
    Sum = Sum + (i)*p[i]
    mu[i] = Sum

# total image mean
mu_t = mu[-1]

#computing between class variance
var_b = ((mu_t*omega - mu)**2)/(omega*(1-omega))

# identifying T where variance is maximum
T = np.nanargmax(var_b)

print('Thresh = ', T)
```

In the above snippet, initially the image probability distribution is computed from the histogram. Afterwards, using for loop, class probability and class mean are calculated according to equations (3.3) and (3.4). Similarly, total mean and between class variance are computed as per equations (3.5) and (3.6). The NumPy library function nanargmax() is used to identify the brightness level where between class variance is maximum. The argmax() function provides the location of maximum value in array. As the array might contain NaN (not a number, used to represent result that is undefined) values, nanargmax () is used to identify the location of maximum variance. This function ignores *NaN* values present in the array. The value of T computed for the present image is 105.

```
img_arr = np.asarray(img)
# apply threshold
thresh_img = np.where(img_arr<=T,0,255)
# image visualization
plt.imshow(thresh_img,cmap='gray')
```

The image is segmented by using the threshold computed in the previous snippet. To apply a threshold, the where() function is used from the NumPy library. It will identify all the pixels that are less than or equal to the threshold and replace them with 0. Similarly, the rest of the pixels would be replaced by 255. The resultant image visualization is performed using the Matplotlib library. The segmented image is

Figure 3.2. Image after thresholding.

presented in figure 3.2. The image separates the foreground elements, which are flowers, from the leafy background.

The above snippets explain step-by-step implementation of threshold computation. There is an inbuilt function in OpenCV threshold(), which implements image thresholding using Otsu's technique. It can threshold the image by calculating the threshold using Otsu's scheme. The implementation is present in the following snippet, and the output is present in figure 3.3. The threshold values obtained by the inbuilt function is 105 that is the same as obtained in the above step-by-step process.

```
T, thresh_img1 = cv2.threshold(img, 0, 255, cv2.THRESH_BINARY + cv2.THRESH_OTSU)
print(T)
plt.imshow(thresh_img1,cmap='gray')
plt.axis('off')
```

In the above implementations, only one threshold is obtained that divides an image in two groups. The Otsu's technique and other thresholding-based techniques can be used to identify multiple thresholds in an image and divide it into multiple parts. This technique is known as multilevel thresholding [8, 9, 19]. Similarly, in the above code a grayscale image was considered, but this scheme could be used in color images too [20]. A color image that a human can perceive consists of three components: Red, Green, and Blue (RGB). A simple color image segmentation algorithm applies the thresholding mechanism separately to RGB components.

Figure 3.3. Thresholded image obtained by OpenCV inbuilt function.

There are different colorspaces in which a color image can be represented like HSV, CIELab, YUV, etc [20]. There are different image segmentation algorithms that perform segmentation tasks in these colorspaces [21].

3.4 Edge detection

Edge detection is one of the most common operations used in image processing [2, 22]. In images, edges are nothing but the boundaries of the objects present in the image. The edge pixels in the image are identified by abrupt change in pixel intensities. This abrupt change is identified as object boundary. There are different types of edge detection models like Roberts, Prewitt, Sobel, etc [2]. These are gradient operators or kernels which are convolved with the image to provide edges of the object present in the image. In equations (3.8) and (3.9) Prewitt and Sobel kernels are shown.

$$P_x = \begin{bmatrix} -1 & 0 & 1 \\ -1 & 0 & 1 \\ -1 & 0 & 1 \end{bmatrix}, P_y = \begin{bmatrix} -1 & -1 & -1 \\ 0 & 0 & 0 \\ 1 & 1 & 1 \end{bmatrix} \tag{3.8}$$

$$S_x = \begin{bmatrix} -1 & 0 & 1 \\ -2 & 0 & 2 \\ -1 & 0 & 1 \end{bmatrix}, S_y = \begin{bmatrix} -1 & -2 & -1 \\ 0 & 0 & 0 \\ 1 & 2 & 1 \end{bmatrix} \tag{3.9}$$

The P_x and P_y are Prewitt and S_x and S_y are Sobel operators or kernels. The two matrices are used to identify vertical and horizontal edges in the images. The image is filtered using these kernels resulting in two images with vertical and horizontal edges I_x and I_y. The complete edge image is obtained using equation (3.10).

$$I_{edge} = |I_x| + |I_y|\qquad(3.10)$$

These operators are very simple operators and provide good results by detecting object edges in the image. Another very famous edge detection technique is the Canny edge detection algorithm [2]. It is a multistep process to obtain clean edges from an image. In the following part, Python implementation of edge detection is discussed. Initially Prewitt's and Sobel's operators are demonstrated using spatial filtering method. Later, inbuilt function in OpenCV for Sobel and Canny edge detection are described. The packages required for edge detection task are same as used during thresholding-based technique.

```
# read image file 1.5.jpg
img = cv2.imread('images\\1.5.jpg')

# convert color image to grayscale
img = cv2.cvtColor(img,cv2.COLOR_BGR2GRAY)

# resize image 256x256
img = cv2.resize(img,(256,256))

plt.imshow(img,cmap = 'gray')
```

A different image is used to show working of edge-based segmentation technique. The image is present in figure 3.4.

```
# generate Prewitt kernel
prewitt_y = np.array([[-1,-1,-1],[0,0,0],[1,1,1]])
prewitt_x = np.array([[-1,0,1],[-1,0,1],[-1,0,1]])
print(prewitt_y)
print(prewitt_x)

# generate Sobel kernel
sobel_y = np.array([[-1,-2,-1],[0,0,0],[1,2,1]])
sobel_x = np.array([[-1,0,1],[-2,0,2],[-1,0,1]])
print(sobel_y)
print(sobel_x)
```

In the above snippet, horizontal and vertical kernels are generated for Prewitt and Sobel edge detection models. To obtain edges the image is filtered using a 2D image convolution operator. To apply these filters on an image, the filter2D() function is

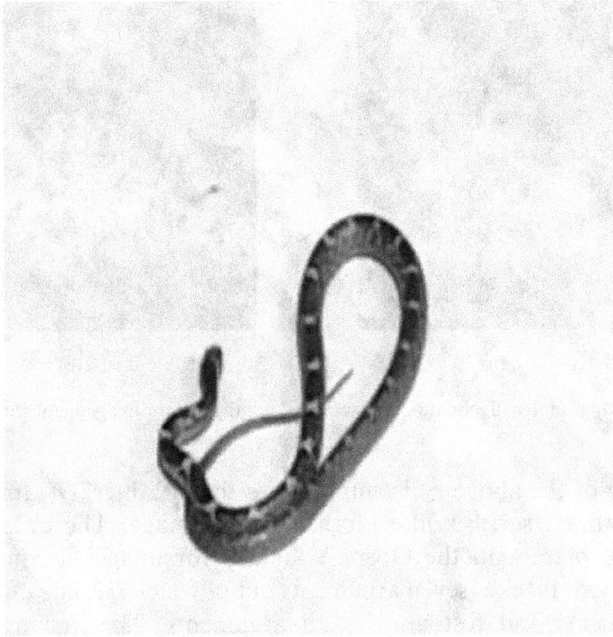

Figure 3.4. Grayscale image used for edge detection, it is output of above code snippet.

used from the OpenCV library. The filtering process is present in the following code snippet.

```python
# spatial filtering image using sobel_x operator
out_imgx = cv2.filter2D(src = img, ddepth = -1,kernel = sobel_x)

# image visualization using matplotlib
plt.subplots(figsize = (12,5))
plt.subplot(1,2,1)
plt.imshow(out_imgx,cmap='gray')

# spatial filtering image using sobel_y operator
out_imgy = cv2.filter2D(src = img,ddepth = -1,kernel = sobel_y)
# image visualization using matplotlib
plt.subplot(1,2,2)
plt.imshow(out_imgy,cmap='gray')
plt.axis('off')

out_imgx = np.asarray(out_imgx,dtype=np.float32)
out_imgy = np.asarray(out_imgy,dtype=np.float32)

out_img = np.abs(out_imgx) + np.abs(out_imgy)
plt.imshow(out_img,cmap='gray')
```

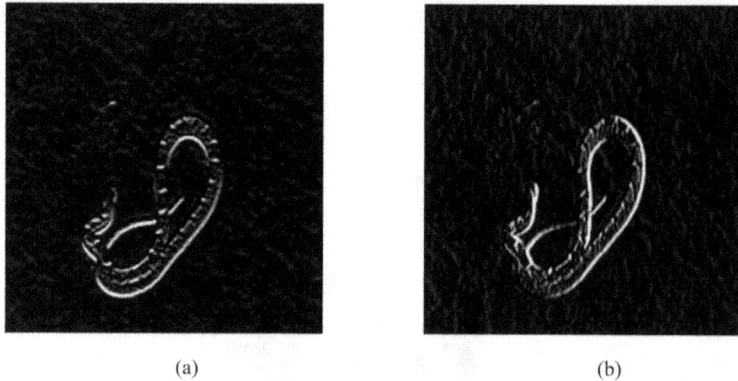

(a)   (b)

Figure 3.5. (a) Output image highlighting horizontal edges. (b) Output image highlighting vertical edges.

The first line of the above code snippet uses the cv2.filter2D() function to apply previously generated sobel_*x* filter kernel on the image. The cv2.filter2D() is an inbuilt function present in the OpenCV library for image filtering using a convolution operation. It takes seven arguments out of which *src*, *ddepth*, and *kernel* are required arguments and rest are default arguments. The first argument to the function is the input image, the second argument is bit resolution or bit depth of the output image, which is kept as −1 to make output bit depth same as that of the input image. The last argument is a kernel for filtering the image. The output is stored in out_imgx variable and visualized by imshow(). The output image after application of the horizontal filter is present in figure 3.5(a). The image is prominently showing the horizontal profile of the snake. The vertical Sobel filter is also applied to the image and the output image after application of the vertical filter is present in figure 3.5(b). The image is clearly highlighting the vertical profile.

The output images are added according to equation (3.10). The resultant image with both vertical and horizontal edges highlighted is present in figure 3.6.

```
# OpenCV inbuilt Sobel edge detection
ox = cv2.Sobel(img,-1,1,0)
oy = cv2.Sobel(img,-1,0,1)

out = np.abs(ox) + np.abs(oy)

plt.imshow(out,cmap='gray')
```

The cv2.Sobel() function is an inbuilt OpenCV function that accepts nine arguments, out of which four are required arguments. The first argument is the input image, the second argument is depth as discussed in the above function. The third and fourth arguments define the order of the filter kernel. The output image is present in figure 3.7, and is similar to image presented in figure 3.6.

A similar kind of output is obtained by using a Prewitt operator on the image. The Prewitt and Sobel operators are simple gradient-based operators. To obtain a

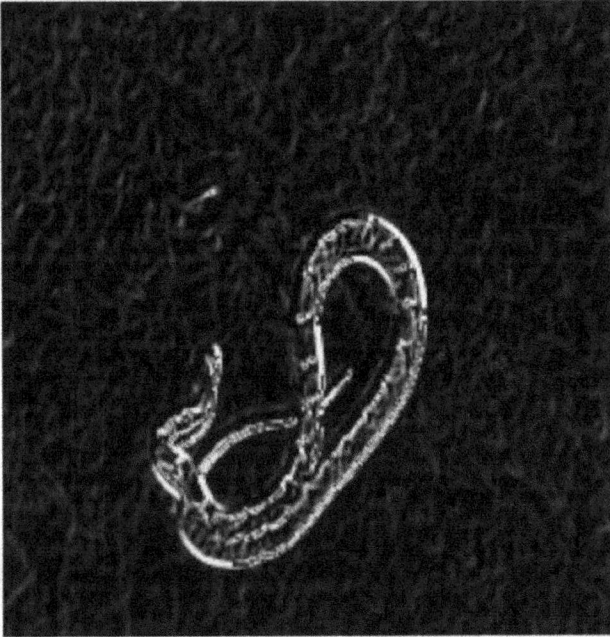

Figure 3.6. Sobel edge detection.

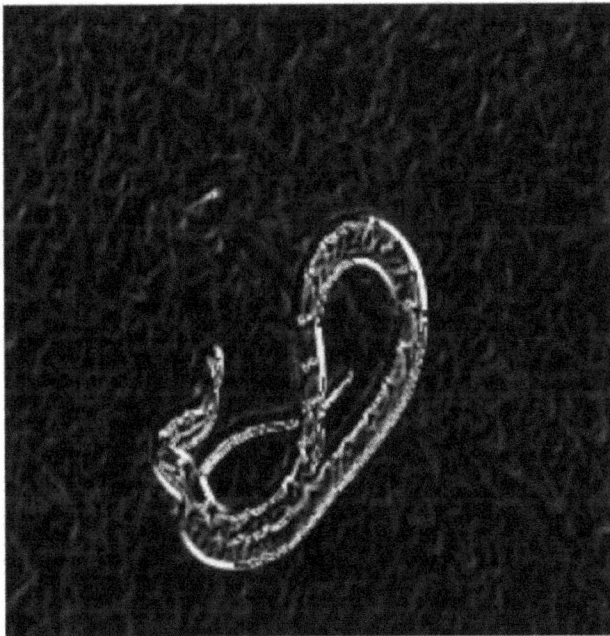

Figure 3.7. OpenCV inbuilt Sobel edge detection output.

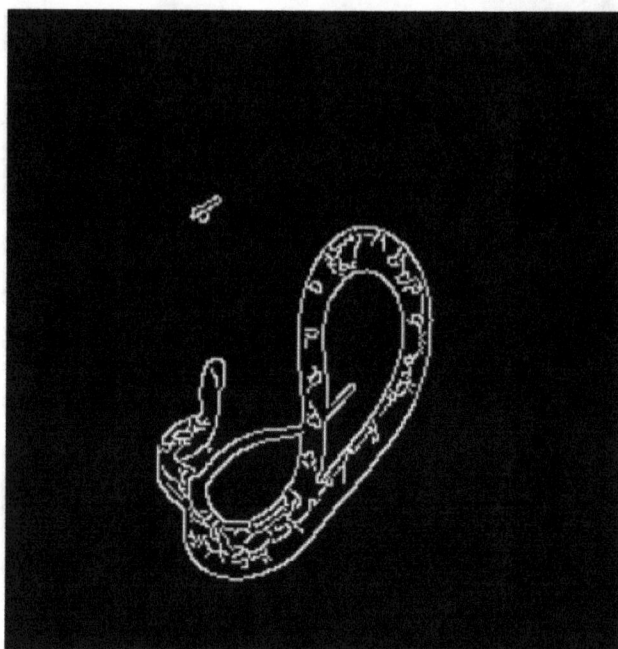

Figure 3.8. Output of Canny edge detection.

superior edge detection performance, the Canny edge detection algorithm is used. The Canny algorithm is a complex algorithm as compared to above methods; it consists of number of steps to obtain edges from the image. The first step is a Gaussian filtering to denoise the image. The output of the Gaussian filtering step is a smoothed version of the input image. The second step is image filtering using gradient-based kernels. The last steps are non-maximum suppression and thresholding. The non-maximum suppression is used to thin out the edges by selecting a single pixel that has maximum value from a neighborhood of the pixels. The last step is double thresholding, which is used to detect and connect the broken edges. The Python implementation of the Canny edge detection algorithm is present in following code snippet. The three arguments present in the function denotes the input image, first threshold and second threshold. The output of the Canny algorithm is present in figure 3.8.

```
# inbuilt Canny edge detection function
o_img = cv2.Canny(img,0,255)
plt.imshow(o_img,cmap='gray')
```

It is clear from the above image that the performance of Canny's algorithm is superior as compared to other techniques. It provides binary image with thin edges. Edge detection has application in computer vision, pattern recognition, etc [23]. In the next section K-means clustering-based image segmentation is discussed.

3.5 Machine learning based image segmentation

This section focuses on the application of machine learning techniques for image segmentation [24]. For image segmentation, supervised machine learning methods are commonly used. These techniques require annotated training data, which includes input images and their corresponding ground truth segmentations. By learning from labeled examples, these methods can predict segmentation boundaries or labels for new images. Popular algorithms for supervised image segmentation includes Random Forests (RF) [25], Support Vector Machines (SVMs) [26, 27], and deep learning networks like convolutional neural networks (CNNs) [24]. The machine learning segmentation technique discussed in this chapter is based on K-Means clustering which is an unsupervised learning method.

3.6 Unsupervised and semi-supervised machine learning approaches

Methods like unsupervised and semi-supervised machine learning are designed to segment images without relying on manually labeled training data. These approaches analyze the image's inherent patterns, structures, or similarities. Clustering algorithms, such as fuzzy c-mean clustering [28, 29] and K-means, are particularly useful for unsupervised image segmentation. In this section, discussion of image segmentation using a clustering technique called K-means clustering (KMC) is present. The concepts behind K-means clustering have been explored and its strengths and limitations have also been discussed. The KMC algorithm is useful in situations where labeled training data is limited or not available.

3.7 Clustering algorithm using KMC

The KMC algorithm is a famous unsupervised learning method for locating unique classes or clusters within a data. Unlike supervised algorithms, it does not rely on labeled data for training. Instead, it groups data points based on their similarity. The number of clusters to be detected in this approach is indicated by the value of k. KMC is a common solution in many applications because of its popularity and efficiency. The flowchart of the process describing the KMC technique is present in figure 3.9.

The algorithm gradually converges into a stable solution by iteratively updating the assignments and recalculating the cluster centroids. The final result is a set of k clusters, where the data points within each cluster exhibit higher similarity than those in different clusters.

It must be understood that the choice of initial cluster centroids can affect the outcome of the algorithm, as different initializations may lead to different solutions. Additionally, the algorithm can be terminated based on convergence criteria, such as when the assignments and centroid updates no longer significantly change.

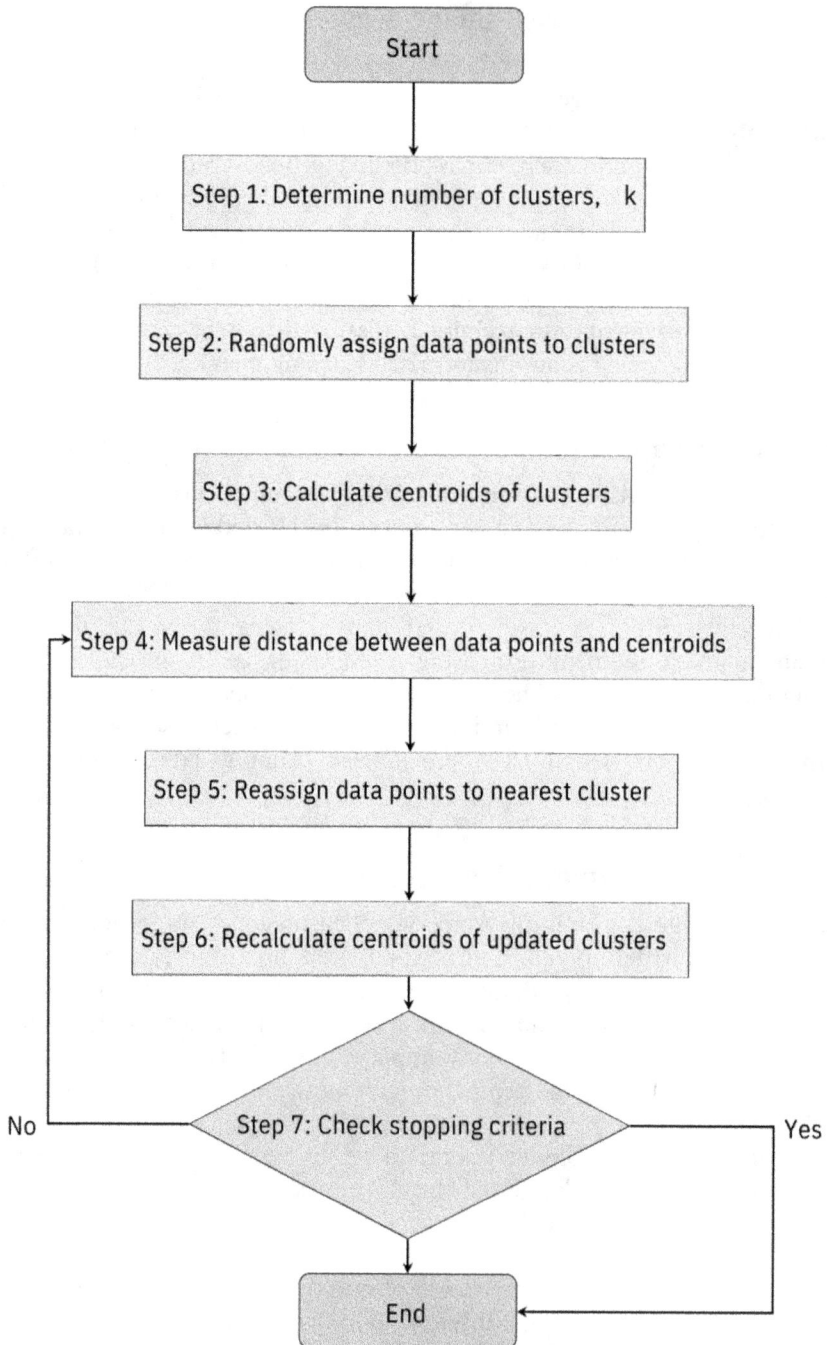

Figure 3.9. Process of K-means clustering.

The KMC algorithm provides a straightforward approach to partitioning data into distinct groups based on similarity. Its efficiency and versatility have made it popular for various clustering tasks, including image segmentation.

3.8 Implementation using Python

Before proceeding with the implementation, it must be ensured that the following dependencies are installed on the system.

```
import cv2
import matplotlib.pyplot as plt
import numpy as np
from sklearn.cluster import KMeans
from skimage.metrics import peak_signal_noise_ratio as psnr
from skimage.metrics import structural_similarity as ssim
```

The implementation uses NumPy, which provides essential numerical computing capabilities. Matplotlib, which is a popular visualization library, has been used for visualizing data. OpenCV, a powerful computer vision library, is used for reading the images. Once all the necessary packages are installed the image is read and pre-processed. To implement K-means clustering, the sklearn library is used. To assess the quality of segmented images for different numbers of clusters, two metrics - peak-signal to noise ratio (PSNR) and Structural SIMilarity (SSIM) - have been used [8]. A higher PSNR value is desired in the output images. Similarly, the SSIM value lies between 0 and 1, the higher the better. The sklearn library is famous for various data analysis tools. The K-means clustering function as well as metrics functions have been used from the sklearn library. The description of the sklearn KMeans function is present in the following part.

```
# Reading the image
img = cv2.imread('images\\3.jpg')

# Convert the image to RGB for proper visualization
img = cv2.cvtColor(img, cv2.COLOR_BGR2RGB)

# Displaying the loaded image
plt.imshow(img)
```

The cv2.imread() function reads the image, and cv2.cvtColor() performs a conversion of the image color space from the default BGR (Blue-Green-Red) to the RGB (Red-Green-Blue) color space. Finally, plt.imshow() displays the RGB image using Matplotlib plotting capabilities. It must be ensured that the image file is available at the specified path. After executing the code, the output RGB image is present in figure 3.10.

Figure 3.10. Displaying the image.

To apply K-means clustering on the image, some data preparation steps are required. Because the image is a three-dimensional shape (height, width, and color channels), the data needs to be reshaped into a two-dimensional array so that it may be processed by the K-means method. The process involves converting the 3D image representation into a flattened 2D array, where each pixel is treated as a separate data point. This transformation allows application of K-means clustering.

By reshaping the image into a 2D array, the spatial information of the image can be effectively represented in a format suitable for the clustering algorithm. This data preparation step is crucial as it helps to leverage the power of K-means clustering for segmenting the image based on similarities in pixel attributes.

```
# Transforming the Image to a 2D Pixel Array with RGB color channels
reshaped_img = np.array(img.reshape((-1,3)),np.float32)
```

`image.reshape()` is used to reshape the image into a 2D array. The use of '−1' within the reshape function serves as a convenient placeholder, allowing the function to automatically determine the appropriate size of the flattened dimension based on the other dimensions of the image. Implementing the K-means algorithm to segment an image is the next step. Initially segmentation is performed for $k = 3$, so the algorithm will find three clusters in the image.

```
# using KMeans from sklearn for obtaining labels and clusters
kmeans = KMeans(n_clusters=3, random_state=10, n_init="auto", max_iter=100,
tol = 0.0001).fit(reshaped_img)

# final centroids are present in kmeans.cluster_centers_
# the returned centroid is float value, converting it to uint8
clusters = np.uint8(kmeans.cluster_centers_)

#output labels are of the size of input, converting it to image shape
labels = kmeans.labels_.reshape(img.shape[:2])

# generating segmented image
segmented_image = clusters[labels]

#image visualization
plt.imshow(segmented_image)

#Computing metrics
print('PSNR = '+f"{psnr(img,segmented_image):.2f}")
print('SSIM = '+f"{ssim(img,segmented_image,channel_axis = -1):.2f}")
```

Starting with $k = 3$, the data will be divided in three clusters. The KMeans() function present in sklearn takes in eight arguments. All the arguments are default arguments i.e., just by providing input data clustering they can be performed. The first argument provided is the number of clusters, its value is 3 here and its default value is 8. The centroids are initialized randomly at the start. The second argument *random_state* $= 10$ ensures the same set of random numbers and initializations in multiple runs of the program. The next argument *n_init* $=$ 'auto' automatically decides the number of initial runs of the method using different seed value for centroids. The *max_iter* and *tol* defines the number of iterations for single run of the algorithm and tolerance. The tolerance is signifying the change in cluster centers in consecutive runs, and when the change in the centroid is less than the tolerance the algorithm stops even if iterations are left.

The labels produced as the output of the clustering algorithm are then modified to take the shape of the original image. The Matplotlib function plt.imshow() is used to display the segmented image. The image present in figure 3.11 is the segmented image obtained after K-means clustering. The image is divided into three clusters, which is evident from the image as it has only three colors: yellow, blackish, and gray. The yellow color in the image is the portion of the image that has yellow-colored flowers. Similarly, the darker portion of the image are leaves and the background. The PSNR of the segmented image with the original image is 19.73 and the SSIM is 0.63.

In figure 3.12, output of the K-means clustering algorithm is present for $k = 6$. The resulting output shows the segmented image with 6 recognized clusters. The PSNR of the segmented image with the original image is 23.61 and the SSIM is 0.69.

Figure 3.11. Segmented image for $k = 3$.

Figure 3.12. Segmented image for $k = 6$.

By increasing the value of k in the K-means algorithm, the resulting segmented image becomes more distinct and clear. This improvement occurs because the algorithm can classify more color clusters or classes. The PSNR and SSIM also increases with increase in number of clusters. The PSNR and SSIM values for different values of k are present in table 3.1. In figure 3.13, the segmented image for $k = 16$ is present, and is showing an image created using 16 colors.

K-means clustering is particularly effective when dealing with smaller datasets, as it can successfully segment objects in images and produce more accurate outcomes.

Table 3.1. PSNR and SSIM comparison for different values of k.

k	PSNR	SSIM
3	19.73	0.63
6	23.61	0.69
8	25.28	0.73
10	26.32	0.76
16	28.11	0.80

Figure 3.13. Segmented image for $k = 16$.

The K-means clustering algorithm is termed a lazy learning algorithm. It means that in this algorithm model is not created separately, instead it works on complete data whenever result is required. Therefore, K-means clustering becomes computationally expensive when applied to larger datasets with more images. This is primarily due to the algorithm examining all samples in a single iteration, significantly increasing processing time.

Therefore, while K-means clustering can yield impressive results for image segmentation, it is essential to consider the computational implications when working with larger datasets.

3.9 Summary

Segmentation is a very important area of research in digital image processing. It is one of the central image processing tasks as it is one of the major steps in high-level image processing algorithms. Segmentation is primarily used to extract important areas or features from the images. In this chapter a comprehensive overview of different image segmentation techniques and their practical implementation is discussed. Image segmentation, its importance, applications, and its different types are discussed. The chapter focuses on three key techniques: thresholding-based image segmentation, edge detection algorithms, and machine learning-based image segmentation.

The first technique discussed is thresholding-based segmentation. The chapter explores Otsu's scheme, a widely used method for automatic threshold selection. The chapter explores three commonly used edge detection algorithms: Prewitt, Sobel, and Canny edge detection. The chapter explores machine learning techniques for image segmentation, focusing specifically on the K-Means clustering algorithm. The chapter offers a comprehensive and practical overview of image segmentation techniques using Python. It equips readers with the necessary knowledge and code implementations to apply thresholding, edge detection, and machine learning algorithms for image segmentation tasks.

References

[1] Pal N R and Pal S K 1993 *A review on image segmentation techniques Pattern Recognit.* **26** 1277–94

[2] Woods R E and Gonzalez R C 2008 *Digital Image Processing* (London: Pearson)

[3] Hesamian M H, Jia W, He X and Kennedy P 2019 *Deep learning techniques for medical image segmentation: achievements and challenges J. Digit. Imaging* **32** 582–96

[4] Sharma N, Ray A K, Shukla K K, Sharma S, Pradhan S, Srivastva A and Aggarwal L 2010 *Automated medical image segmentation techniques J. Med. Phys./Assoc. Med. Phys. India* **35** 3

[5] Liu X, Song L, Liu S and Zhang Y 2021 A review of deep-learning-based medical image segmentation methods *Sustainability* **13** 1224–4

[6] Yuan K, Zhuang X, Schaefer G, Feng J, Guan L and Fang H 2021 Deep-learning-based multispectral satellite image segmentation for water body detection *IEEE J. Sel. Top. Appl. Earth Obs. Remote Sens.* **14** 7422–34

[7] Pare S, Bhandari A K, Kumar A, Singh G K and Khare S 2015 Satellite image segmentation based on different objective functions using genetic algorithm: a comparative study *IEEE Int. Conf. on Digital Signal Processing (DSP)* (IEEE: Singapore) 730–4

[8] Kumar A, Kumar A, Vishwakarma A and Singh G K 2022 *Multilevel thresholding for crop image segmentation based on recursive minimum cross entropy using a swarm-based technique Comput. Electron. Agric.* **203** 107488

[9] Kumar A, Kumar A, Vishwakarma A and Lee H-N 2022 *An improved segmentation technique for multilevel thresholding of crop image using cuckoo search algorithm based on recursive minimum cross entropy IET Signal Proc.* **16** 630–49

[10] Hao S, Zhou Y and Guo Y 2020 *A brief survey on semantic segmentation with deep learning Neurocomputing* **406** 302–21

[11] Liu X, Deng Z and Yang Y 2019 *Recent progress in semantic image segmentation Artif. Intell. Rev.* **52** 1089–106

[12] Guo Y, Liu Y, Georgiou T and Lew M S 2018 *A review of semantic segmentation using deep neural networks Int. J. Multimed. Inf. Retr.* **7** 87–93

[13] Li Z and Chen J 2015 *Superpixel segmentation using linear spectral clustering 2015 IEEE Conf. on Computer Vision and Pattern Recognition (CVPR)* (Boston, MA: IEEE) 1356–63

[14] Achanta R, Shaji A, Smith K, Lucchi A, Fua P and Süsstrunk S 2012 *SLIC superpixels compared to state-of-the-art superpixel methods IEEE Trans. Pattern Anal. Mach. Intell.* **34** 2274–81

[15] Van Den Bergh M, Boix X, Roig G, De Capitani B and Van Gool L 2012 *SEEDS: superpixels extracted* via *energy-driven sampling Lecture Notes in Computer Science* (Berlin: Springer) vol 7578 13–26

[16] Ban Z, Liu J and Cao L 2018 Superpixel segmentation using gaussian mixture model *IEEE Trans. Image Process.* **27** 4105–17

[17] Otsu N 1979 *Threshold selection method from gray-level histograms IEEE Trans. Syst. Man Cybern.* **SMC-9** 62–6

[18] Kohler R 1981 *A segmentation system based on thresholding Comput. Graph. Image Process.* **15** 319–38

[19] Pare S, Kumar A, Bajaj V and Singh G K 2019 *A context sensitive multilevel thresholding using swarm based algorithms IEEE/CAA J. Autom. Sin.* **6** 1471–86

[20] Cheng H D, Jiang X H, Sun Y and Wang J 2001 *Color image segmentation: advances and prospects Pattern Recognit.* **34** 2259–81

[21] Bai X D, Cao Z G, Wang Y, Yu Z H, Zhang X F and Li C N 2013 Crop segmentation from images by morphology modeling in the CIE L*a*b* color space *Comput. Electron. Agric.* **99** 21–34

[22] Marr D and Hildreth E 1980 *Theory of edge detection Proc. R Soc. Lond.* B **207** 187–217

[23] Zhai L, Dong S and Ma H 2008 Recent methods and applications on image edge detection *2008 Int. Workshop on Education Technology and Training and 2008 Int. Workshop on Geoscience and Remote Sensing, ETT and GRS 2008* **1** 332–5

[24] Ghosh S, Das N, Das I and Maulik U 2019 Understanding deep learning techniques for image segmentation *ACM Comput. Surv. (CSUR)* **52** 1–35

[25] Yang T, Song J and Li L 2019 *A deep learning model integrating SK-TPCNN and random forests for brain tumor segmentation in MRI Biocybern. Biomed. Eng.* **39** 613–23

[26] Li S, Fevens T, Krzyzak A and Li S 2006 *Automatic clinical image segmentation using pathological modeling, PCA and SVM Eng. Appl. Artif. Intell.* **19** 403–10

[27] Yang H Y, Wang X Y, Wang Q Y and Zhang X J 2012 *LS-SVM based image segmentation using color and texture information J. Vis. Commun. Image Represent.* **23** 1095–112

[28] Aja-Fernández S, Curiale A H and Vegas-Sánchez-Ferrero G 2015 *A local fuzzy thresholding methodology for multiregion image segmentation Knowl. Based Syst.* **83** 1–12

[29] Jiao J, Wang X, Wei T and Zhang J 2023 *An adaptive fuzzy c-mean noise image segmentation algorithm combining local and regional information IEEE Trans. Fuzzy Syst.* **31** 2645–57

Chapter 4

Segmentation of digital images with region growing algorithm

Amin Sakhaei and Zahra Ghanbari

Image segmentation plays a pivotal role in image processing. With the advent and proliferation of artificial intelligence (AI), and its various algorithms deployed in digital image processing, assessing the accuracy of these algorithms has become increasingly important. However, manual segmentation is a labor-intensive and time-consuming task that requires considerable expertise and knowledge. To mitigate these issues, a traditional technique known as 'region growing' is often utilized in image segmentation. Region growing is a method that consolidates pixels or subregions into larger regions, following specific growth criteria.

This study aimed to improve the segmentation of different regions of the images. Therefore, a pre-processing such as adaptation was performed on additional images, and then a region-growing algorithm was performed using eight-neighborhood to generate new pixels. Finally, different areas were separated from each other.

4.1 Introduction

Segmentation has always been one of the main problems and issues in digital image processing. Also, splitting images into different parts is considered one of the main topics in computer vision. Therefore, the need to use simple methods with reliable results that can compare the efficiency of the effects of more complex algorithms with them felt very essential. Image segmentation is used in various fields of medicine, self-driving cars, object detection, pedestrian detection, recognition parts of satellite images, recognition tasks (face, fingerprint, …), video surveillance, etc.

Image segmentation algorithms primarily fall into two categories:

(1) **Similarity-based:** These algorithms are rooted in the premise that a specific object's intensity values are similar across its background. Techniques within this category include histogram-based methods such as K-means

doi:10.1088/978-0-7503-5924-5ch4

clustering and thresholding, Otsu's method, and strategies like region-growing and merging.

(2) **Discontinuity (dissimilarity)-based:** This approach partitions an image into various regions by exploiting existing discontinuities. During the process of edge detection, the object's edges within the image create discontinuities. These discontinuities allow the object to be distinguished from the remainder of the image [1].

This study uses 'Region Growing' as the main algorithm. It will be discussed in detail below.

4.2 Pre-processing methods for image segmentation

To use the region-growing algorithm, the image must be pre-processed first. Pre-processing is the same in some cases, such as conversion to grayscale. However, other pre-processing depends on the shooting conditions, such as image quality, the intensity of colors, the quality of the image, etc.

4.2.1 Converting to grayscale

Experimental results showed that segmentation in grayscale images is more accurate than RGB images. Below are the reasons for converting images into grayscale in segmentation.

- **Simplification:** Grayscale images are comprised solely of shades of gray, usually depicted as 8-bit pixel values that range from 0 (representing black) to 255 (representing white). This representation simplifies the image data when compared with RGB images, which incorporate three color channels—red, green, and blue. Each of these channels uses 8 bits, thus constituting 24 bits per pixel. By converting images to grayscale, we can decrease both the storage and processing demands associated with image data.
- **Reduce complexity of algorithms:** many image processing algorithms are designed to work with grayscale images. These algorithms can be simpler and faster when applied to grayscale images compared to color images. Grayscale images serve as an effective starting point for various image processing tasks, including image enhancement, segmentation, and object detection.
- **Retaining essential image information:** Grayscale conversion retains the essential information of an image while discarding color information. For some applications, such as image analysis for medical, scientific, or technical purposes, color information may not be relevant or necessary. Grayscale images can effectively represent an image's intensity or luminance information, often crucial for image analysis tasks.
- **Human perception:** Grayscale images are often used in applications focusing on human perception, such as medical imaging or artistic photography. Grayscale images can convey important information or evoke specific emotions without the distraction of color, making them effective in particular contexts.

Grayscale conversion is commonly used in image processing due to its simplicity, efficiency, compatibility, and suitability for specific image analysis tasks. It is a valuable technique for simplifying and processing image data, and it has a wide range of applications in various fields [2, 3].

4.2.2 Resize

In some cases, the recorded image has a considerable length and width, increasing the calculation volume. For this reason, the image size should be reduced to the desired level.

4.2.3 Histogram equalization

Many researchers affirm that histogram equalization is a straightforward method to adjust an image's contrast and brightness. This adjustment is accomplished by redistributing the pixel intensity values across the entire dynamic range of the image. The primary goal of histogram equalization is to expand the intensity levels of an image to utilize the full range of possible intensities. This expansion results in an image with an enhanced visual appearance and detailed features [4].

Histogram equalization operates by mapping the original intensity values of the image to new ones using a transformation function. This function is dictated by the cumulative distribution function (CDF) of the image's original histogram. The CDF represents the cumulative probability of a pixel's intensity being less than or equal to a certain level within the image.

The process of histogram equalization proceeds as follows:
1. Compute the cumulative distribution function (CDF) of the histogram.
2. Normalize the CDF to span the full intensity range (0–255 in an 8-bit image).
3. Using the CDF transfer function to generate new intensities.
4. Generate image using new intensities (The result is depicted in Figure 4.1).

Histogram equalization is a straightforward and frequently utilized method in image processing. It enhances the contrast in images suffering from uneven or low contrast, including those with overly dark or light regions. However, it doesn't always yield the best results, as it can also intensify noise and lead to over-enhancement in some situations. Various adaptations of histogram equalization, like adaptive histogram equalization, can help overcome some of these basic limitations [2].

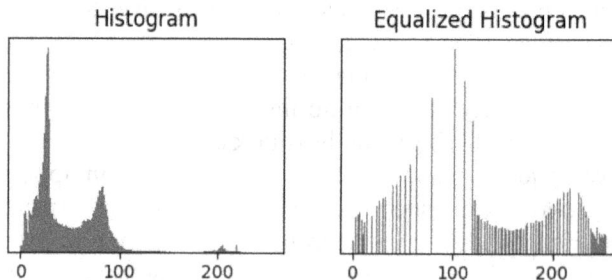

Figure 4.1. Histogram equalization.

4.2.4 Edge detection

Edge detection, a fundamental technique in image processing, involves pinpointing and emphasizing the edges or boundaries of objects or regions in an image. It serves as a preliminary step in various computer vision and image analysis tasks such as object recognition, image segmentation, and feature extraction.

In an image, edges are characterized by areas of quick intensity transitions or discontinuities, where the intensity values shift abruptly. Edge detection algorithms are designed to identify these rapid changes in intensity, which often represent object boundaries, lines, corners, or other structural elements within the image.

Edge detection algorithms generally function on grayscale images, but they can also be adapted to color images after first converting them to grayscale. Image processing utilizes a variety of well-known edge detection techniques, such as:

1. **Sobel operator:** The Sobel operator, a commonly utilized gradient-based edge detection technique, calculates the gradient of the image intensity at each pixel using a collection of convolution kernels, typically in horizontal and vertical orientations. The edge's strength is represented by the gradient's magnitude, while its orientation is indicated by the gradient's direction. It can be shown as $\begin{bmatrix} w_{11} & w_{12} & w_{13} \\ w_{21} & w_{22} & w_{23} \\ w_{31} & w_{32} & w_{33} \end{bmatrix}$, where we have:

$$s_x = \frac{\partial f}{\partial x} = (w_{31} + 2w_{32} + w_{33}) - (w_{11} + 2w_{12} + w_{13}) \tag{4.1}$$

$$s_y = \frac{\partial f}{\partial y} = (w_{13} + 2w_{23} + w_{33}) - (w_{11} + 2w_{21} + w_{31}) \tag{4.2}$$

$$\begin{aligned} M(x, y) \approx |(w_{31} + 2w_{32} + w_{33}) - (w_{11} + 2w_{12} + w_{13})| \\ + |(w_{13} + 2w_{23} + w_{33}) - (w_{11} + 2w_{21} + w_{31})| \end{aligned} \tag{4.3}$$

2. **Canny edge detector:** Renowned for its accurate and reliable results, the Canny Edge Detector is a prevalent edge detection algorithm. It leverages a multi-stage process encompassing Gaussian smoothing, gradient computation, non-maximum suppression, and hysteresis thresholding. This process helps in detecting edges while minimizing false positives and negatives. The method is graphically represented in figure 4.2.

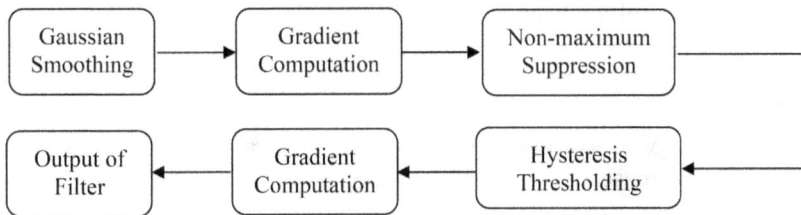

Figure 4.2. Canny edge detector steps [5].

3. **Roberts operator:** The Roberts operator is another gradient-based edge detection technique that uses a pair of simple 2×2 convolution kernels to compute the gradient in the diagonal directions. It is computationally simple but may be less accurate than other methods in some cases.

$$\begin{bmatrix} w_{11} & w_{12} \\ w_{21} & w_{22} \end{bmatrix}$$

$$s_x = w_{22} - w_{11} \tag{4.4}$$

$$s_y = w_{21} - w_{12} \tag{4.5}$$

$$M(x, y) = \sqrt{(w_{22} - w_{11})^2 + (w_{21} - w_{12})^2} \tag{4.6}$$

4. **Prewitt operator:** Resembling the Sobel operator, the Prewitt operator employs subtly different convolution kernels to calculate the gradient in both horizontal and vertical directions. As a gradient-based technique, it too can be utilized for edge detection.

$$\begin{bmatrix} w_{11} & w_{12} & w_{13} \\ w_{21} & w_{22} & w_{23} \\ w_{31} & w_{32} & w_{33} \end{bmatrix}$$

$$s_x = \frac{\partial f}{\partial x} = (w_{31} + w_{32} + w_{33}) - (w_{11} + w_{12} + w_{13}) \tag{4.7}$$

$$s_y = \frac{\partial f}{\partial y} = (w_{13} + w_{23} + w_{33}) - (w_{11} + w_{21} + w_{31}) \tag{4.8}$$

$$\begin{aligned} M(x, y) \approx &|(w_{31} + w_{32} + w_{33}) - (w_{11} + w_{12} + w_{13})| \\ &+ |(w_{13} + w_{23} + w_{33}) - (w_{11} + w_{21} + w_{31})| \end{aligned} \tag{4.9}$$

Edge detection plays a pivotal role in numerous image processing tasks by offering crucial insights about the borders and structures of objects or areas within an image. The results from an edge detection algorithm often serve as input for subsequent image processing or computer vision algorithms, enabling extraction of significant features or higher-level analysis [2, 6].

4.2.5 Smoothing filter

In the realm of image processing, a technique known as a smoothing filter or blurring filter is used to decrease noise or intricate details in an image, thereby creating a smoother image rendition. It operates by calculating the average of pixel values within a specified neighborhood or window in the image, then replaces the central pixel with this computed average. The degree of applied image smoothing is governed by the size of this neighborhood or window, often referred to as the kernel size.

Various image processing tasks such as image denoising and image enhancement commonly employ smoothing filters. These filters aid in diminishing the impacts of noise or minor image inconsistencies, while also softening sharp edges and fine details. Different mathematical operations like convolution or averaging can be used to apply smoothing filters, and the type of kernel used—Gaussian, median, or mean, for example—is determined by the specific application and desired level of smoothing. The selection of an appropriate smoothing filter hinges on the image characteristics and the particular image processing task at hand.

Averaging (box filter): The box filter, commonly referred to as a 'nearest-neighbor' filter, functions by selecting the value of the closest pixel within a window and assigning it to the output pixel. This filter is utilized as a low-pass filter to diminish the image resolution while retaining its overall structure. The box filter is typically applied by dividing the image into non-overlapping rectangular regions, called blocks or tiles, and replacing the pixels within each block with the average value or the value of the nearest pixel, depending on the specific application.

The box filter is a simple and fast filter, as it does not require any complex mathematical operations or kernel weights. However, it may result in loss of fine details and introduce aliasing artifacts, especially while down sampling an image with high-frequency content. Therefore, other more advanced techniques such as bilinear or bicubic interpolation are often used for image down-sampling when better image quality is desired. The choice of the appropriate filter and method for image down-sampling depends on the specific application and the desired level of image quality.

$$\frac{1}{n} \times \begin{bmatrix} 1 & 1 & 1 \\ 1 & 1 & 1 \\ 1 & 1 & 1 \end{bmatrix}$$

The choice of the appropriate filter and kernel size should be based on the specific image processing task and desired level of smoothing [2].

Example:

$$\begin{bmatrix} 136 & 209 & 242 & 16 & 128 \\ 196 & 64 & 233 & 189 & 4 \\ 50 & 226 & 35 & 52 & 177 \\ 142 & 2 & 27 & 138 & 119 \\ 66 & 102 & 147 & 58 & 9 \end{bmatrix} \times \left(\frac{1}{9} \begin{bmatrix} 1 & 1 & 1 \\ 1 & 1 & 1 \\ 1 & 1 & 1 \end{bmatrix} \right) = \begin{bmatrix} x & x & x & x & x \\ x & 145 & x & x & x \\ x & x & x & x & x \\ x & x & x & x & x \\ x & x & x & x & x \end{bmatrix}$$

$$\frac{38}{9} + \frac{136}{9} + \frac{209}{9} + \frac{134}{9} + \frac{196}{9} + \frac{64}{9} + \frac{252}{9} + \frac{50}{9} + \frac{226}{9} = 145$$

4.2.6 Zero padding

Zero padding is a technique employed in image processing and other signal-processing applications. It aims to expand the dimensions of an image or signal by appending zeros (pixels or samples with zero values) along its borders. This is done to create a larger image or signal that has the same spatial dimensions as the original, but with additional 'padding' or 'border' of zeros.

Zero padding is commonly used in image processing operations that involve filtering or convolution, such as edge detection, image sharpening, or image blurring. When applying a filter or convolution operation to an image, the filter typically 'slides' or 'convolves' over the image by computing a weighted sum of pixel values within a neighborhood or window centered at each pixel. Zero padding allows the filter to be applied to the pixels at the edges of the image as well, by extending the image with zeros so that the filter can fully cover the entire image, including its edges.

The primary objective of zero padding is to prevent information loss at the edges of an image during filtering or convolution processes. When padding is not employed, the filter solely covers a portion of the pixels located at the center of the image. Consequently, the resulting output image would be smaller compared to the input image. This disparity can introduce artifacts and distortions at the edges of the image since the filter weights might not be entirely applied to the edge pixels. As a consequence, the outcomes may be incomplete or inaccurate. Zero padding helps to mitigate these issues by providing additional information at the edges of the image, allowing the filter to be applied more accurately across the entire image.

$$
\begin{bmatrix} x & x & x & x & x \\ x & x & x & x & x \\ x & x & x & x & x \\ x & x & x & x & x \\ x & x & x & x & x \end{bmatrix} \xrightarrow{Zero\ padding} \begin{bmatrix} 0 & 0 & 0 & 0 & 0 & 0 \\ 0 & x & x & x & x & 0 \\ 0 & x & x & x & x & 0 \\ 0 & x & x & x & x & 0 \\ 0 & x & x & x & x & 0 \\ 0 & 0 & 0 & 0 & 0 & 0 \end{bmatrix}
$$

4.3 Image segmentation methods

Image segmentation involves the task of partitioning an image into distinct and meaningful regions or segments that exhibit coherence and relevance. There are several different methods for image segmentation, each has its own advantages, limitations, and areas of application. Some common image segmentation methods are:

- **Thresholding:** Thresholding is a simple and widely used image segmentation method. It involves setting a threshold value and classifying pixels or regions based on their intensity values with respect to this threshold. Pixels or regions with intensity values exceeding the specified threshold are categorized as one segment, whereas those below the threshold are classified as a separate segment. Thresholding is effective when the object of interest has distinct intensity values compared to the background.

- **Edge-based methods:** Edge-based methods rely on the identification of sudden changes in intensity values, which indicate the presence of boundaries or edges between distinct regions within an image. Popular techniques for edge detection include Sobel, Canny, and Roberts operators. Once the edges are detected, additional processing steps can be carried out to connect or refine the edges, ultimately forming coherent regions.

- **Region-based methods:** Region-based methods encompass the task of grouping pixels or regions that exhibit similar characteristics, such as intensity values, color, texture, or other image properties. These methods typically initiate the process with a seed region and progressively expand it by incorporating neighboring pixels or regions that satisfy specific criteria or rules. Prominent examples of region-based methods include region growing, region merging, and watershed segmentation.

- **Clustering methods:** Clustering methods are employed to group pixels or regions based on their similarity in a feature space. Common clustering techniques comprise k-means clustering, fuzzy c-means clustering, and hierarchical clustering. These methods are suitable for unsupervised image segmentation, where the number of segments or regions is not predetermined.

- **Machine learning-based methods:** Machine learning-based methods leverage algorithms and models to acquire the segmentation task by learning from a labeled training set of images. Examples encompass supervised methods like support vector machines (SVMs), and convolutional neural networks (CNNs). These methods exhibit high accuracy and can effectively handle intricate image segmentation tasks.

- **Graph-based methods:** Graph-based methods represent an image as a graph, where pixels or regions are nodes, and edges represent relationships or connections between neighboring pixels or regions. Graph-based methods use graph theory concepts and algorithms, such as minimum spanning tree, normalized cut, and random walker, to segment the image into regions.

- **Active contour models:** Active contour models, also known as snakes, are energy-based methods that use deformable curves or contours to represent object boundaries. The contour evolves over time to minimize an energy function that is defined based on image characteristics, such as intensity gradients or region statistics. Active contour models are often used for object extraction or boundary refinement tasks.

The aforementioned methods represent a selection of image segmentation techniques within the realm of image processing and computer vision. However, these are merely a few illustrations, as there are numerous other methods available. The selection of a suitable method hinges upon the specific requirements of the application, the image's characteristics, and the desired outcomes. It is common practice to combine or customize different methods to cater to a particular image segmentation task [1, 2, 7–9].

4.4 The proposed method

Region growing is an extensively utilized image segmentation algorithm in the field of image processing. Its purpose is to group together pixels or regions with similar characteristics, such as intensity values, color, texture, or other image properties. This algorithm belongs to the category of region-based image segmentation, which aims to partition an image into regions or regions of interest (ROIs) that exhibit shared properties.

The region-growing algorithm initiates by selecting an initial 'seed' pixel or region. It then progressively expands the region through iterative steps by incorporating neighboring pixels or regions that meet specific criteria or predefined rules. The criteria for region growth can vary depending on the particular application and the desired characteristics of the segmented regions. For example, in a basic intensity-based region-growing algorithm, pixels or regions with intensity values falling within a certain range relative to the seed pixel or region are included within the expanding region. The region growing process continues until no further pixels or regions meeting the criteria can be added or until a predefined stopping condition is met.

Region growing algorithms find applications in a variety of image processing tasks, including object or feature detection, image segmentation, and image analysis. They are particularly useful for segmenting regions with homogeneous properties, such as regions with similar intensity values or regions with similar texture patterns. Region growing algorithms are relatively simple to implement and can be applied to both grayscale and color images. However, they may be sensitive to the choice of seed pixel or region, and the criteria for region growth, which may require careful tuning to achieve desired results. Additionally, region-growing algorithms may suffer from issues such as over-segmentation or under-segmentation, depending on the image characteristics and the specific algorithm used. Therefore, the proper understanding of the image properties and careful parameter selection are important when using region-growing algorithms for image processing tasks [2, 3].

$$In\ 8-bit\ image: \begin{cases} 0 & \textit{if the absolute difference between} \\ & \textit{seed and the pixel} \leqslant \textit{Thresholld} \\ 255 & \textit{Otherwise} \end{cases} \qquad (4.10)$$

Here is a visual representation of the steps in the algorithm we are proposing, shown in figure 4.3.

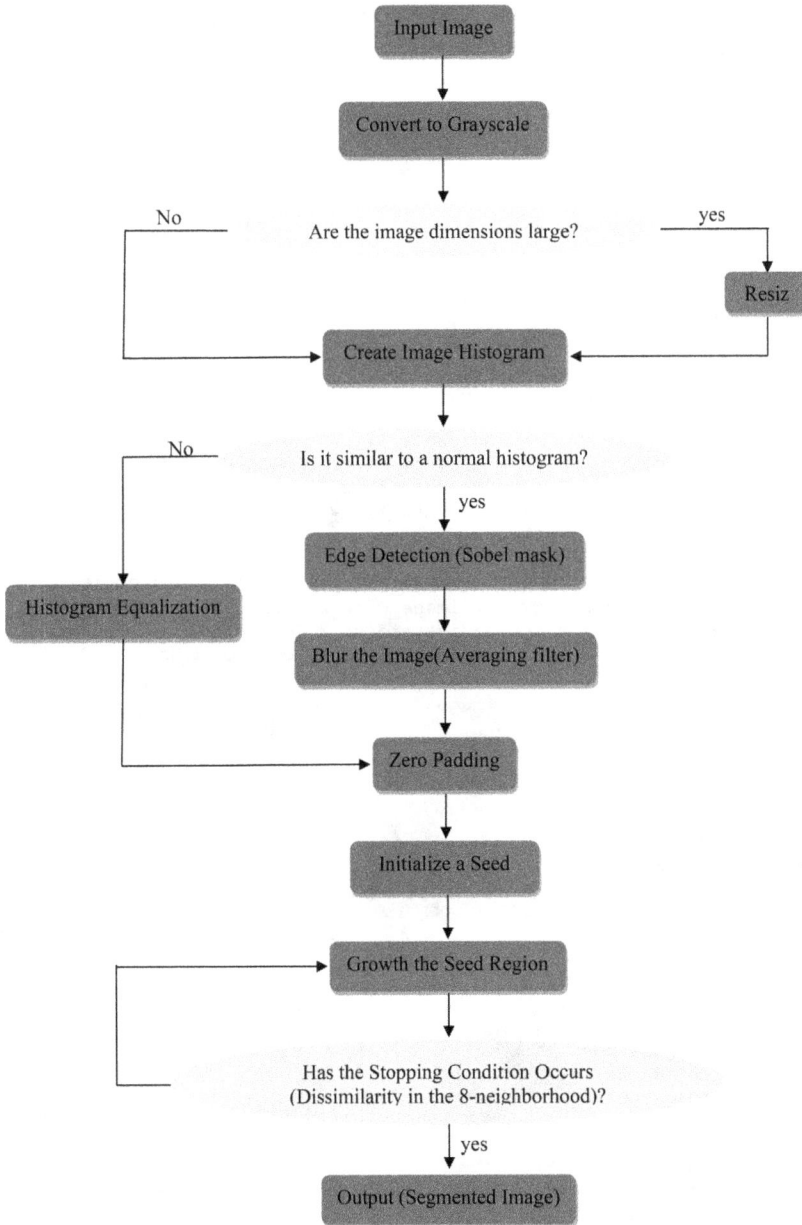

Figure 4.3. Proposed algorithm flowchart.

4.4.1 Pre-processing Python code (resize and histogram equalization)

In this section, the Python code for the region growth algorithm is thoroughly explained in each subsection. The four images are segmented by a seed that is done by clicking on that part, the desired area is segmented, and finally the results are observed. In most cases, all parts of the code are the same for all images. Where there is a difference, it will be noted.

4.4.1.1 *Converting images to grayscale*

```python
# Load the required libraries
import cv2
import numpy as np
import matplotlib.pyplot as plt

# Load the images into system memory
image_data_1 = cv2.imread("1.jpeg")
image_data_2 = cv2.imread("2.jpg")
image_data_3 = cv2.imread("3.jpg")
image_data_4 = cv2.imread("4.jpg")

# Converting to RGB
image_data_1 = cv2.cvtColor(image_data_1, cv2.COLOR_BGR2RGB)
image_data_2 = cv2.cvtColor(image_data_2, cv2.COLOR_BGR2RGB)
image_data_3 = cv2.cvtColor(image_data_3, cv2.COLOR_BGR2RGB)
image_data_4 = cv2.cvtColor(image_data_4, cv2.COLOR_BGR2RGB)

# Converting to grayscale
image_data_1_gray = cv2.cvtColor(image_data_1, cv2.COLOR_BGR2GRAY)
image_data_2_gray = cv2.cvtColor(image_data_2, cv2.COLOR_BGR2GRAY)
image_data_3_gray = cv2.cvtColor(image_data_3, cv2.COLOR_BGR2GRAY)
image_data_4_gray = cv2.cvtColor(image_data_4, cv2.COLOR_BGR2GRAY)

# Plot original and grayscale images
plt.figure()

ax_s_s = plt.subplot(2, 4, 1)
ax_s.set_title('House')
ax_s.imshow(image_data_1)
ax_s.axis('off')
ax_s = plt.subplot(2, 4, 2)
ax_s.set_title('Egg')
ax_s.imshow(image_data_2)
ax_s.axis('off')
ax_s = plt.subplot(2, 4, 3)
ax_s.set_title('Flower')
ax_s.imshow(image_data_3)
ax_s.axis('off')
ax_s = plt.subplot(2, 4, 4)
ax_s.set_title('Snake')
ax_s.imshow(image_data_4)
ax_s.axis('off')

ax_s = plt.subplot(2, 4, 5)
ax_s.imshow(image_data_1_gray, cmap='gray', vmin=0, vmax_s=255)
ax_s.axis('off')
ax_s = plt.subplot(2, 4, 6)
ax_s.imshow(image_data_2_gray, cmap='gray', vmin=0, vmax_s=255)
ax_s.axis('off')
ax_s = plt.subplot(2, 4, 7)
ax_s.imshow(image_data_3_gray, cmap='gray', vmin=0, vmax_s=255)
ax_s.axis('off')
```

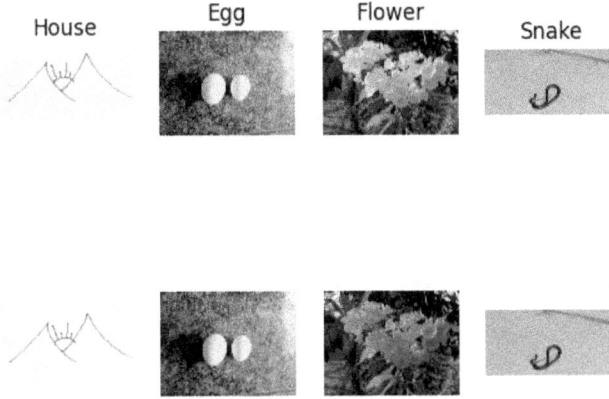

Figure 4.4. Grayscale image.

```
House Image Dimension (542, 997)
Egg Image Dimension (240, 320)
Flower Image Dimension (240, 320)        Resized Snake Image Dimensions (309, 654)
Snake Image Dimension (1854, 3926)
```

Figure 4.5. (a) Images dimension. (b) Resized image dimension.

```
ax_s = plt.subplot(2, 4, 8)
ax_s.imshow(image_data_4_gray, cmap='gray', vmin=0, vmax_s=255)
ax_s.axis('off')
# For display figures (It is usually placed at the end of the code, but it
is placed in this section so that there is no problem in the implementation
of each section)
plt.show()
```

The output of the above-shown code is represented in figure 4.4.

4.4.1.2 Image dimensions

```
# Print dimensions
print(image_data_1_gray.shape)
print(image_data_2_gray.shape)
print(image_data_3_gray.shape)
print(image_data_4_gray.shape)
```

The output of the above-shown code is represented in figure 4.5(a). Accordingly, the snake image (image_data_4_gray) should be resized.

4.4.1.3 Resize

```
# Resizing snake image
# Dimension reduction
scale = 6  # Percent of original size
width = int(image_data_4_gray.shape[1] / scale)
height = int(image_data_4_gray.shape[0] / scale)
dim = (width, height)

# Resize Image
image_data_4_resized = cv2.resize(image_data_4_gray, dim,
interpolation=cv2.INTER_AREA)
print(image_data_4_resized.shape)
```

The output of the above-shown code is represented in figure 4.5(b).

4.4.1.4 Histogram

```
# Plot images histogram
plt.figure()

ax_s = plt.subplot(2, 2, 1)
ax_s.hist(image_data_1_gray.ravel(), bins=256,  range=(0, 255))
ax_s.set_title('House')
ax_s.ax_ses.get_yaxis().set_visible(False)
ax_s = plt.subplot(2, 2, 2)
ax_s.hist(image_data_2_gray.ravel(), bins=256,  range=(0, 255))
ax_s.set_title('Egg')
ax_s.ax_ses.get_yaxis().set_visible(False)
ax_s = plt.subplot(2, 2, 3)
ax_s.hist(image_data_3_gray.ravel(), bins=256,  range=(0, 255))
ax_s.set_title('Flower')
ax_s.ax_ses.get_yaxis().set_visible(False)
ax_s = plt.subplot(2, 2, 4)
ax_s.hist(image_data_4_resized.ravel(), bins=256,  range=(0, 255))
ax_s.set_title('Snake')
ax_s.ax_ses.get_yaxis().set_visible(False)

# Add space between subplots
plt.subplots_adjust(left=0.1, bottom=0.1, right=0.7, top=0.7, wspace=0.5,
hspace=0.5)
plt.show()
```

The output of the code displayed above is depicted in figure 4.6. As shown in figure 4.4, the image of the snake (image_data_4_resized) bears no resemblance to the standard histogram. Thus, it is necessary to perform histogram equalization.

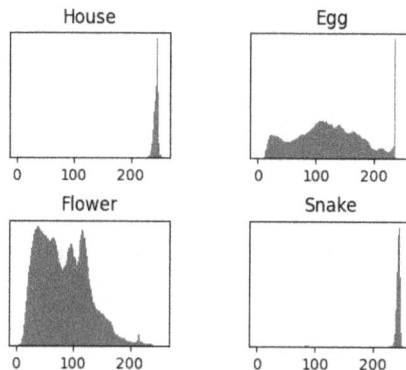

Figure 4.6. Histograms.

4.4.1.5 Histogram equalization

```python
# Vectorize Histogram
def vhistogram(x):
    dicts = {}
    v = np.zeros(256)
    for p in range(x.shape[0]):
        for q in range(x.shape[1]):
            v[x[p, q]] = v[x[p, q]]+1
            dicts[x[p, q]] = v[x[p, q]]
    return v, dicts

# Cumulative distribution function (CDF) calculator
def cdf_func(c, dicts):
    for p in range(255):
        c[p] = c[p] + c[p-1]
        dicts[p] = c[p]
    return c, dicts

# Function to normalize CDF
def normalize(image_data):
    l = 256
    m, n = image_data.shape
    vhisto, dicts = vhistogram(image_data)
    cdf, dicts = cdf_func(vhisto, dicts)
    h = np.zeros(l)
    for p in range(l-1):
        h[p] = ((cdf[i]-cdf.min())/((m*n)-cdf.min()))*(l-1)



        h[p] = h[p].astype('uint8')
        dicts[p] = h[p]
    for p in range(m):
        for q in range(n):
            image_data[p, q] = dicts.get((image_data[p, q]))
    return image_data

image_data_4_copy = image_data_4_resized.copy()
image_data_4_HE= normalize(image_data_4_copy)

# Plot results
plt.figure()

ax_s = plt.subplot(2, 2, 1)
ax_s.set_title('Snake')
ax_s.imshow(image_data_4_resized, cmap='gray', vmin=0, vmax_s=255)
ax_s.axis('off')
ax_s = plt.subplot(2, 2, 2)
ax_s.set_title('Snake Improved Contrast')
ax_s.imshow(image_data_4_HE, cmap='gray', vmin=0, vmax_s=255)
ax_s.axis('off')
ax_s = plt.subplot(2, 2, 3)
ax_s.hist(image_data_4_resized.ravel(), bins=256,  range=(0, 255))
ax_s.set_title('Histogram')
ax_s.ax_ses.get_yaxis().set_visible(False)
ax_s = plt.subplot(2, 2, 4)
ax_s.hist(image_data_4_HE.ravel(), bins=256)
ax_s.set_title('Equalized Histogram')
ax_s.ax_ses.get_yaxis().set_visible(False)

cv2.imwrite('4_Equalized_Histogram.jpg', image_data_4_HE)

plt.show()
```

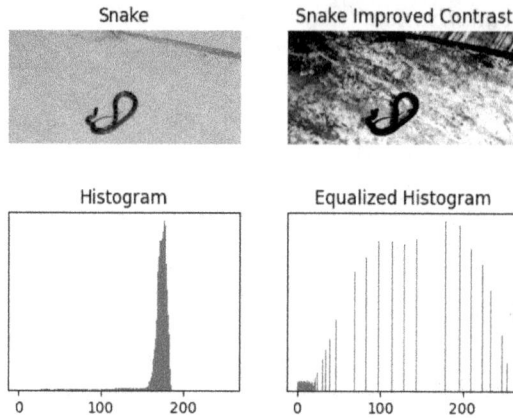

Figure 4.7. Equalized histogram (snake).

Initially, we compute the abundance vector of the image. Subsequently, employing the cumulative distribution function (CDF), we derive the volume distribution function. Finally, we map the resultant solution from 0 to 255 by iterating through each intensity. The conversion of the normalized frequency vector is done as follows: In the first function, we define a dictionary with image pixel intensities and their frequencies. In the CDF function, we convert the abundance to the volume distribution for each intensity and replace it with the new intensity of the image. Finally, we use a loop to produce an image with updated intensities. The output of the above-shown code is represented in figure 4.7.

4.4.2 Pre-processing Python code (edge detection and blurring)

4.4.2.1 Edge detection

```python
# Load the image into system memory
image_data = plt.imread('1.jpg',0)

# Edge detection
# Define filters for detecting vertical and horizontal edges
vert_filter = [[-1,-2,-1], [0,0,0], [1,2,1]]
horiz_filter = [[-1,0,1], [-2,0,2], [-1,0,1]]

# Initialize the edges image
edges = image_data.copy()

# Loop over all pixels in the image
for p in range(3, image_data.shape[0] - 2):
    for q in range(3, image_data.shape[1] - 2):

        # Extract a 3x3 patch of pixels centered at the current pixel
        patch = image_data[p - 1:p + 2, q - 1:q + 2, 0]

        # Compute edge scores using the vertical and horizontal filters
        vert_score = np.sum(patch * vert_filter) / 4
        horiz_score = np.sum(patch * horiz_filter) / 4

        # Combine the vertical and horizontal scores into a total edge
score
        score = np.sqrt(vert_score ** 2 + horiz_score ** 2)

        # Insert the edge score into the output image
        edges[p, q] = [score] * 3 plt.figure()
plt.imshow(edges, cmap='gray', vmin=0, vmax_s=255)
plt.axis('off')
```

House

House Improved Contrast

Histogram

Equalized Histogram

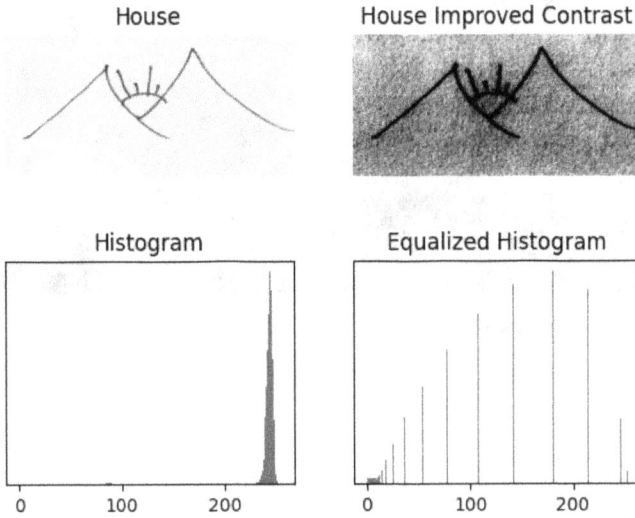

Figure 4.8. Equalized histogram (house).

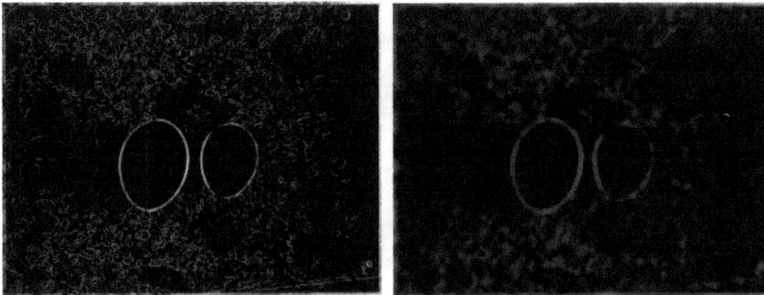

Figure 4.9. (a) Edge detected image. (b) Blurred image.

This code performs edge detection on a grayscale image using vertical and horizontal filters. It loops through each pixel in the image, extracts a local 3×3 box of pixels centered at the current pixel, and applies the filters to calculate vertical and horizontal edge scores. These scores are combined using the Euclidean distance formula to obtain a total edge score, which is then stored in a new 'edges image' as grayscale values representing the strength of edges in the original image. The output of the above-shown code is represented in figure 4.8(a). The same process should be applied to the other two images (egg and flower) as well. The results are shown in figures 4.9(a) and 4.10(a) for simplicity.

$$\textit{Sobel Vertical Filter:} \begin{bmatrix} -1 & -2 & -1 \\ 0 & 0 & 0 \\ 1 & 2 & 1 \end{bmatrix}, \textit{Sobel Horizontal Filter:} \begin{bmatrix} -1 & 0 & 1 \\ -2 & 0 & 2 \\ -1 & 0 & 1 \end{bmatrix} [7]$$

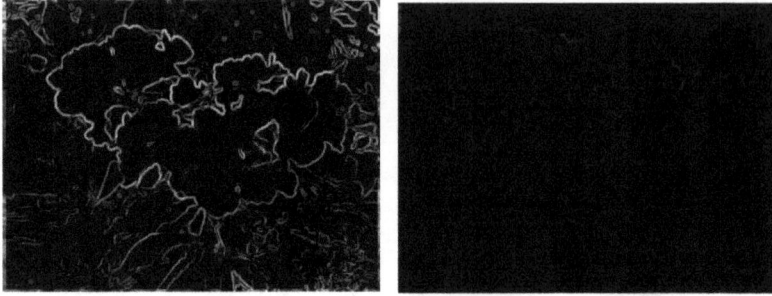

Figure 4.10. (a) Edge detected image. (b) Blurred image.

4.4.2.2 Blurring image

```python
# Define a kernel for blurring
kernel = np.ones((4, 4),np.float32)/50

# Apply the filter2D function to perform convolution and blur the edges
image
dst = cv2.filter2D(edges_image_data,-1,kernel)

plt.figure()
plt.imshow(dst, cmap='gray', vmin=0, vmax_s=255)
plt.axis('off')

cv2.imwrite('PreProcessed.jpg', dst)

plt.show()
```

This code blurs the edges image ('edges_image_data_1') using a 4×4 kernel and displays the resulting blurred image using Matplotlib. The 'cv2.filter2D()' function is used to apply the kernel for blurring, and the blurred image is converted to unsigned integer data type.

The kernel used to blur the image of egg and flower is as follows:

$$Egg: \frac{1}{25} \times \begin{bmatrix} 1 & 1 & 1 & 1 & 1 \\ 1 & 1 & 1 & 1 & 1 \\ 1 & 1 & 1 & 1 & 1 \\ 1 & 1 & 1 & 1 & 1 \\ 1 & 1 & 1 & 1 & 1 \end{bmatrix}$$

$$Flower: \frac{1}{25} \times \begin{bmatrix} 1 & 1 & 1 \\ 1 & 1 & 1 \\ 1 & 1 & 1 \end{bmatrix} [7]$$

These values were obtained using empirically. The output of the above-shown code is represented in figures 4.8(b), 4.9(b), and 4.10(b).

4.4.3 Region growing Python code

4.4.3.1 Region growing algorithm

```python
# Load the image into the system memory
image = cv2.imread('PreProcessed.jpg', 0)

# Variable to count the number of clicks
click_count = 0

# Region Growing Function
def segment(event, x, y, flags, params):
    # Define global variables
    global new_pixels, new_image_data, image_data, seed_intensity, click_count

    image_data = image.copy()

    # Zero padding
    image_data = cv2.copyMakeBorder(image_data, 1, 1, 1, 1, cv2.BORDER_CONSTANT, None, value = 0)

    # To prevent overflow
    image_data = np.int64(image_data)

    ix, iy = y, x

    # Define the array to use later
    new_pixels = np.zeros_like(image_data)
    ze = np.zeros_like(image_data)
    final_image_data = np.zeros_like(image)

    if event == cv2.EVENT_LBUTTONDOWN:
        click_count += 1

        seed_intensity = image_data[ix, iy]

        new_pixels[ix, iy] = 255

        new_image_data = new_pixels.copy()

        # resetting to save the 8-neighborhood new pixels
        new_pixels = np.zeros_like(image_data)
        new_pixels[ix, iy] = 255

        # Region-growing segmentation
        while not np.array_equal(ze, new_image_data):

            ze = new_image_data.copy()
            for p in range(1, image_data.shape[0] - 1):
                for q in range(1, image_data.shape[1] - 1):

                    if new_pixels[p, q] == 255:
                        if abs(image_data[p - 1, q] - seed_intensity) < 11 and new_image_data[p - 1, q] != 255:
                            new_pixels[p - 1, q] = 255
                            new_pixels[p, q] = 0
                            new_image_data[p - 1, q] = 255

                        if abs(image_data[p + 1, q] - seed_intensity) < 11 and new_image_data[p + 1, q] != 255:
                            new_pixels[p + 1, q] = 255
```

```
                             new_pixels[p, q] = 0
                             new_image_data[p + 1, q] = 255

                    if abs(image_data[p, q - 1] - seed_intensity) < 11
and new_image_data[p, q - 1] != 255:
                             new_pixels[p, q - 1] = 255
                             new_pixels[p, q] = 0
                             new_image_data[p, q - 1] = 255

                    if abs(image_data[p, q + 1] - seed_intensity) < 11
and new_image_data[p, q + 1] != 255:
                             new_pixels[p, q + 1] = 255
                             new_pixels[p, q] = 0
                             new_image_data[p, q + 1] = 255

                    if abs(image_data[p - 1, q - 1] - seed_intensity) <
11 and new_image_data[p - 1, q - 1] != 255:
                             new_pixels[p - 1, q - 1] = 255
                             new_pixels[p, q] = 0
                             new_image_data[p - 1, q - 1] = 255

                    if abs(image_data[p + 1, q + 1] - seed_intensity) <
11 and new_image_data[p + 1, q + 1] != 255:
                             new_pixels[p + 1, q + 1] = 255
                             new_pixels[p, q] = 0
                             new_image_data[p + 1, q + 1] = 255

                    if abs(image_data[p - 1, q + 1] - seed_intensity) <
11 and new_image_data[p - 1, q + 1] != 255:
                             new_pixels[p - 1, q + 1] = 255
                             new_pixels[p, q] = 0
                             new_image_data[p - 1, q + 1] = 255

                    if abs(image_data[p + 1, q - 1] - seed_intensity) <
11 and new_image_data[p + 1, q - 1] != 255:
                             new_pixels[p + 1, q - 1] = 255
                             new_pixels[p, q] = 0
                             new_image_data[p + 1, q - 1] = 255

        # Returning the image to uint8
        new_image_data = new_image_data.astype(np.uint8)

        # Remove zero padding
        for p in range(0, image_data.shape[0] - 2):
            for q in range(1, image_data.shape[1] - 2):
                final_image_data[p, q] = new_image_data[p + 1, q + 1]

        if (click_count==1):
            final_image_data_1 = np.zeros_like(final_image_data)
            final_image_data_1 = final_image_data
            cv2.imwrite('Segmented_1.jpg', final_image_data_1)
        if (click_count==2):
            final_image_data_2 = np.zeros_like(final_image_data)
            final_image_data_2 = final_image_data
            cv2.imwrite('Segmented_2.jpg', final_image_data_2)

        cv2.imshow('Segmented Image', final_image_data)
        cv2.waitKey(0)

cv2.namedWindow(winname='Image')
```

Figure 4.11. The result of the snake image.

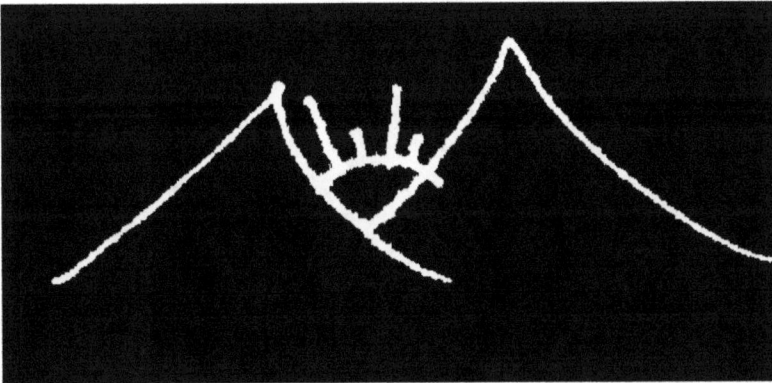

Figure 4.12. The result of the house image.

```
cv2.setMouseCallback('Image', segment)

while True:

    cv2.imshow('Image', image)

    if cv2.waitKey(10) & 0xFF == 27:
        break

cv2.destroyAllWindows()
```

The code performs region-growing segmentation on an image using OpenCV2 in Python. It defines a callback function called 'segment' for mouse events on an image window. When the left mouse button is clicked, the function captures the clicked pixel's intensity, initializes arrays for region growing, and starts the region growing process. It iterates through the image pixels, checks intensity differences, and adds similar pixels to the segmented region. The segmented region is displayed in an OpenCV2 window, and the results are stored. The code also includes a loop to display the original image until the ESC key is pressed.

Coding for two snake and house images ends in this section. The final output of the above-shown code is represented in figures 4.11 and 4.12.

But the remaining two images need to be combined because they include two separate parts.

4.4.4 Combine Python code

4.4.4.1 Combining results

```python
# Load the images into system memory
image_data_s1 = cv2.imread('Segmented_1.jpg', 0)
image_data_s2 = cv2.imread('Segmented_2.jpg', 0)

# Define the array to use later
image_data_final = np.zeros_like(image_data_s1)

# Loop for Combine 2 image
for p in range(image_data_s1.shape[0]):
    for q in range(image_data_s1.shape[1]):
        if (image_data_s1[p, q] > 100):
            image_data_final[p, q] = 255
        elif (image_data_s2[p, q] > 100):
            image_data_final[p, q] = 255

plt.imshow(image_data_final, cmap='gray', vmin=0, vmax_s=255)
plt.axis('off')
plt.show()
```

The code loads two grayscale images ('Segmented_1.jpg' and 'Segmented_2.jpg') using OpenCV2, initializes an array 'image_data_final' to store the combined result, iterates through each pixel of the images, checks the intensity values, and combines them into 'image_data_final' based on certain conditions. Finally, it displays the combined image using Matplotlib with a grayscale colormap and without axis ticks.

The final output of the above-shown code is represented in figures 4.13 and 4.14.

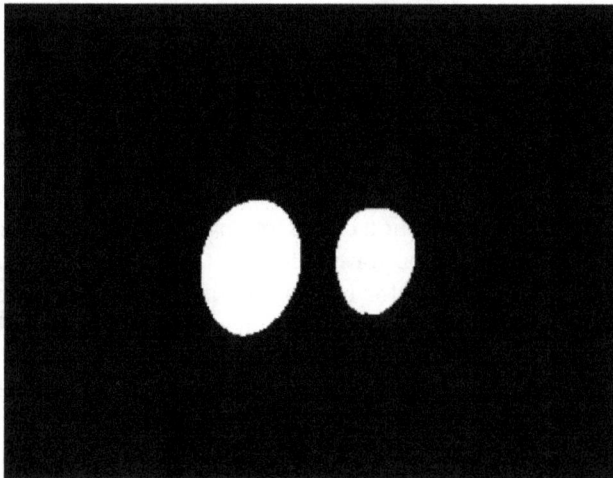

Figure 4.13. The result of the egg image.

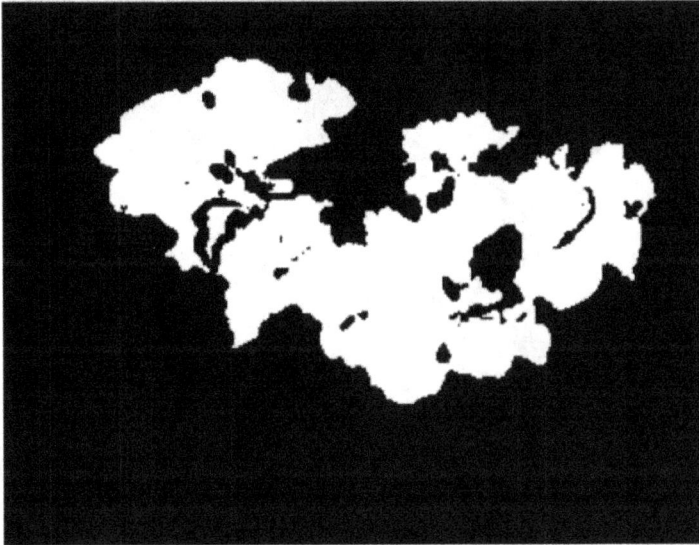

Figure 4.14. The result of flower image.

4.5 Conclusion

The suggested method is a quick one with decent accuracy, but the accuracy heavily relies on choosing the right starting point (mouse click) and knowing the desired area (user control), as illustrated in Figure 4.15. This figure highlights the consequences of selecting an inappropriate seed.

AI-powered algorithms (machine learning) can also do the same task with better accuracy, but they require more complex calculations.

When it comes to labeling medical images, it requires the expertise of an experienced professional. However, due to limitations such as time constraints and fatigue, traditional labeling methods may introduce errors. As a solution, the proposed algorithm can be employed for initial labeling and tasks that do not require high precision. Additionally, the algorithm utilizes advanced techniques like edge detection and histogram equalization for input preprocessing, which enhances its reliability by applying local filters and making adjustments. This allows the algorithm to provide more accurate and dependable results during the pre-processing stage, leading to improved outcomes and greater usability in various medical imaging applications.

This algorithm can separate different regions of the brain, including black and gray matter, in MRI images, which can be used to diagnose diseases such as dementia.

The proposed algorithm was applied in a practical way with the aim of improving the classical methods of image processing and their combination.

To enhance the proposed algorithm in the edge detection section and achieve more distinct edges, it is recommended to explore alternative methods as replacements. These alternative methods can be employed in place of the current approach.

Figure 4.15. The result of selecting the wrong seed.

Acknowledgments

I would like to express my sincere gratitude to Dr Hamed Azarnoush for his invaluable contribution to my understanding of digital image processing during the fall semester of 2020. His passion, expertise, and dedication to his students have greatly influenced my knowledge and skills in this field.

References

[1] Kaur D and Kaur Y 2014 Various image segmentation techniques: a review *Int. J. Comput. Sci. Mob. Comput.* **3** 809–14

[2] Gonzalez R C and Woods R E 2018 *Digital Image Processing* 4th edn (London: Pearson)

[3] Sonka M, Hlavac V and Boyle R 2015 *Image Processing, Analysis, and Machine Vision* 4th edn (Boston, MA: Cengage Learning)

[4] Mustafa W A and Abdul Kader M M M 2018 A review of histogram equalization techniques in image enhancement application *J. Phys.: Conf. Ser.* **1019** 012026

[5] Abdelgawad H M, Safar M and Wahba A M 2015 High Level Synthesis of Canny Edge Detection Algorithm on Zynq Platform *Int. J. Comput. Electr. Autom. Control Inf. Eng.* **9** 148–52

[6] Canny J 1986 A computational approach to edge detection *IEEE Trans. Pattern Anal. Mach. Intell.* **8** 679–98

[7] Arbelaez P, Maire M, Fowlkes C and Malik J 2011 Contour detection and hierarchical image segmentation *IEEE Trans. Pattern Anal. Mach. Intell.* **33** 898–916

[8] Camilus K S and Govindan V K 2012 A review on graph based segmentation *Int. J. Image Graph. Signal Process.* **4** 1–13

[9] Pal N R and Pal S K 1993 A review on image segmentation techniques *Pattern Recognit.* **26** 1277–94

IOP Publishing

Image Processing with Python
A practical approach
Irshad Ahmad Ansari and Varun Bajaj

Chapter 5

Retinal layer segmentation in OCT images

P V Sudeep, V R Deepthi and G Sreelekha

Optical coherence tomography (OCT) is a significant diagnostic tool that provides a cross-sectional view of the retina with high resolution. It enables the identification and assessment of different retinal abnormalities such as age-related macular degeneration, glaucoma and diabetic retinopathy. Such abnormalities spotted at an early stage by evaluating the thickness of different retinal layers, optical nerves, and fluid accumulations can enable prompt treatment and management. Retinal layers need to be segmented accurately before performing quantitative shape analysis. In the literature, different rule-based and learning-based methods are proposed for retinal layer segmentation. Deep learning methods, in particular, have proven to be highly effective in learning segmentation rules automatically from large datasets. In this chapter, we discuss the principles of retinal layer segmentation and elaborate the design cycle of the deep learning methods for retinal layer segmentation in OCT images. To illustrate the design cycle, the U-Net model has been selected and the Python implementation of the same for OCT retinal layer segmentation is presented. The Duke Cyst DME dataset is used for training the model in the experimentation. We also evaluate the experimental results on the test dataset qualitatively and quantitatively.

5.1 Introduction

The eyes are an extraordinary example of the complexity of the human body, with each part working together to transform light into meaningful information that the brain can interpret [1]. Human eyes have the remarkable ability to capture, modify, and convert light into a chemical message that the brain alone can decode.

To reach the iris and pupil, light must first pass through the cornea, which bends it and causes it to converge inside the eye. The iris controls how much light may enter the eye by adjusting the pupil size in response to the amount and quality of light. The lens receives light after it has passed through the pupil and can change its shape with the help of auxiliary muscles in order to focus things at different distances through the process of accommodation. Additionally, the lens enhances the already-clear corneal image before projecting it onto the retina. Through the photoreceptors and pigmented epithelial cells, the retina receives the light. These photoreceptors' photopigment molecules absorb light, which results in electrical signal changes. When light energy is transformed into electrical impulses, a number of signals are set off that pass via the neurons in the retina and then to the optic nerve, which ultimately travels to the brain.

The retina of the human eye plays a crucial role in allowing us to perceive images, as it is the place where the light entering our eye is being processed and translated into neural signals which could be interpreted by the brain. Without the retina, a viewer would be unable to perceive any images. The retina is a crucial neural structure responsible for sensing and processing light information, leading to vision. It consists of various layers with specific functions. The outer region of the retina contains rods and cone cells, which convert light signals into action potentials. These signals are then transmitted to ganglion cells in the inner retina through bipolar neurons. The ganglion cell axons form the optic nerve, which carries the neural signal to the brain for interpretation. The retina also includes the retinal pigment epithelium (RPE), photoreceptor layer, outer nuclear layer (ONL), outer plexiform layer (OPL), inner nuclear layer (INL), ganglion cell layer (GCL), nerve fiber layer (NFL), and the inner limiting membrane (ILM). Each layer plays a role in local processing of the neural signal, allowing for the identification and interpretation of visual stimuli.

For improving the assessment, treatment and management of retinal diseases, high quality retinal imaging is necessary [2]. An optical coherence tomography (OCT) machine generates cross sectional images of tissue in the human eye with high resolution and is valuable in the assessment of different diseases related to the retina. Compared to ultrasound imaging, OCT utilizes light rather than sound and OCT images have higher axial resolution. OCT is fiber optic based and can be associated with a number of medical devices.

Many eye-related abnormalities like glaucoma [3], age-related macular degeneration [4], diabetes-related retinopathy [5], etc., can be diagnosed and managed by OCT analysis [6]. OCT features are useful in the analysis of macular disorders like retinal pigment epithelial detachments, choroidal imaging, central serous chorioretinopathy and choroidal and retinal tumors [7]. By analyzing the thickness of different retinal layers, optical nerves and fluid accumulations, various disease conditions can be diagnosed at an earlier stage. Automated retinal layer segmentation can make this early diagnosis easier. The automatic image analysis method such as segmentation provides quantitative information to clinicians and researchers which aid them in their decision making. Although most commercial OCT devices offer automatic layer analysis, many times they fail to accurately segment all the essential retinal layers automatically [8]. However, the manual and time consuming correction process relies on expertise and is susceptible to bias [9].

As mentioned above, analyzing the distinct features of each retinal layer is essential for making clinical diagnoses because structural changes to the retinal layers are correlated with symptoms of the most prominent eye diseases. Although the OCT provides useful information, additional processing is necessary to figure out data that is therapeutically meaningful. Therefore, segmentation is the fundamental component in image-based eye examination. Millions of OCT procedures are conducted worldwide per year for ocular and neurodegenerative disorders [11]. Thus, the amount and variety of data make manual segmentation not only tiresome but also impractical, which demands efficient automated retinal layer segmentation of OCT images.

5.2 Related works

In this section, a plethora of retinal layer segmentation methods using OCT images are reviewed and an overview of such methods are presented in figure 5.1. It can be broadly classified into rule-based methods [12–14], probabilistic approaches [15, 16] and ML approaches [9, 17–22].

The class of intensity and gradient-based methods for retinal layer segmentation leverage the information contained in the intensity values and gradient of images to extract useful features from images [12–14]. Shahidi *et al* [12] utilized intensity profiles for estimating depth location of peak, which means identifying the points where intensity values are highest. In [13], two dimensional information of gradient and dynamic programming are used to find out the border lines and the thickness of retinal inner layers, such as the IPL, GCL and RNFL to analyze the risk of developing glaucoma. However, the outcomes in segmentation are constrained by retinal layer irregularities and intensity discontinuities. In [14], a 3D weighting function is computed in accordance with gradient and intensity properties of the

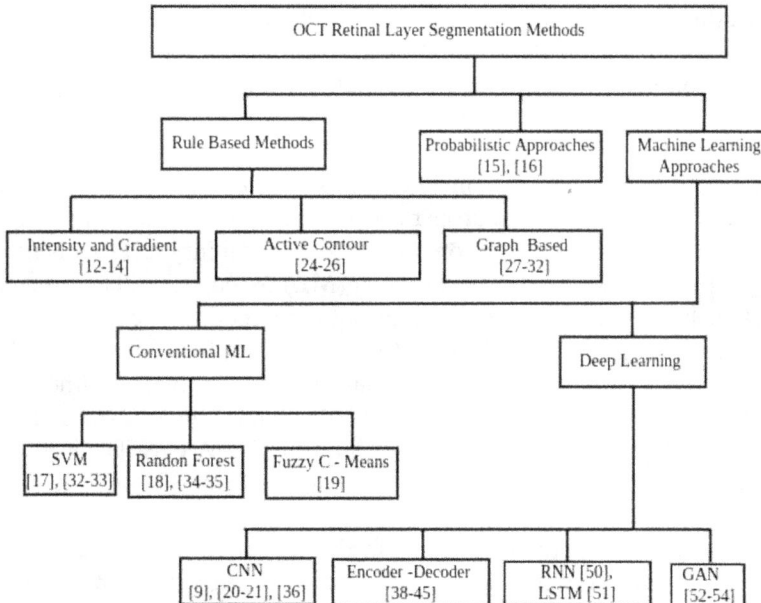

Figure 5.1. Retinal layer segmentation methods.

image. An adaptive thresholding technique has been proposed in [23] which utilized the A-scan reflectivity histogram for fixing cutoffs to locate the border of each layer. After obtaining the border locations, an integrity check was performed to find inconsistencies and abnormalities in it.

Active contour models are effective in capturing the layer boundaries during retinal layer segmentation [24–26]. Fernandez *et al* [24] proposed a segmentation method that uses the local coherence information to identify distinct borders among different retinal layers. The Chan-Vese active contour technique with energy minimization to segment SD-OCT data from mice [25]. Ghorbel *et al* [26] presented a technique for segmenting the eight layers of the retina using global segmentation algorithms such as Markov random fields and active contours.

Furthermore, the graph-based [27–32] methods are also proposed for the same application. The two-step technique in [27] for segmenting retinal layers uses gradient information in dual scales to capture edge details. The method also uses a shortest-path finding method to optimize the edges that have been detected. Automated segmentation methods using SD-OCT images are presented in [28] and [30]. In [29], the method in [28] is used for segmenting seven retinal layers and combining the kernel regression method for identifying the fluid. To reduce the computation complexity, the graph cut-based OCT segmentation method in [32] has been implemented with Dijkstra's algorithm using min-heap.

Different probabilistic approaches are suggested in the literature for retinal layer segmentation [15, 16]. For example, in the user-guided segmentation approach [15], the manually specified border lines at locations, where most of the automatic segmentation approaches are not able to produce correct results due to deformity of retinal layers, are used as a guide by the approach. This method used robust likelihood estimation to detect the whole 3D retinal layer and anatomical details. Also, Sun *et al* [16] suggested a method for segmenting the retinal layers in SD-OCT images using adaptive-guided-coupling-probability level sets. The intensity information for each intraretinal layer is described by a Gaussian fitting distribution for each voxel's probability representation, and the probability map of the boundary that is created by combining adaptive thickness information with anatomical priors to ensure surfaces evolve within the proper range. Then, for detecting layer boundaries, a probability representation based on a coupling probability level set is introduced.

Another approach is the machine learning (ML) based retinal layer segmentation methods which outperformed various existing methods. The conventional ML algorithms such as SVM [17, 32, 33], Random Forest [18, 34, 35] and Fuzzy C-means clustering [19] are some of the popular ones among them. Rather than using sophisticated rules to delineate each layer, elementary features are defined and manually labeled samples are used for training the SVM classifier in [17, 32] and [33]. In [18], eight different retinal layers are detected from 3D OCT images using a random forest classifier. A graph-based segmentation technique proposed by Karri *et al* [34] applied a structured random forest to detect specific borders, and to improve the segmentation. Xiang *et al* [35] also proposed a random forest classifier-based layer segmentation method in which the shapes and intensities of retinal layers are learned using 24 features. The model utilized a modified live-wire approach to identify area among retinal layers accurately, even though the OCT images with fluids have low contrast and the layers are significantly distorted.

Recently, convolutional neural network (CNN)-based techniques [9, 20, 21, 36] are used to segment the retinal layers. In contrast to ML approaches, CNN uses automatically generated features obtained by convolution computations rather than the manually created features. A segmentation method using CNN, trained with nine classes, and a structured random forest classifier has been proposed in literature [9]. Fang *et al* [20] suggested using Graph Search and CNN to segment the nine layers. A network architecture called CNN-S closely inspired by the AlexNet architecture is proposed in [21]. In [36], the method combines features learnt through a deep residual network with manual features to train the random forest classifier. Viedma *et al* [37] investigated various CNN-based architectures for retrieving features from OCT images and classification is performed using different approaches by applying extracted features. The laborious feature selection procedures used in conventional methods have been greatly reduced by this.

It has been shown in literature that the encoder-decoder-based models are achieving better performance in segmentation [38–45]. In [38], authors proposed U-Net architecture with symmetric encoder-decoder layers for segmenting biomedical images. More details on U-Net architecture is provided in the upcoming section of this book chapter. Roy *et al* [39] introduced RelayNet, an encoder-decoder model for retinal layer and fluid segmentation, identifying 7 retinal layers, Region above retina (RaR), Region Below RPE (RbR), and accumulated fluid. In [40], a cascaded fully convolutional network (FCN) [46] used to segment eight retinal layers. The first FCN generated probability maps for the layers, followed by a topology check. Regions failing the topology conditions were processed by a second FCN for topology corrections. In addition, BRU-Net [41] utilized dilated residual blocks in a U-shape configuration, which is asymmetric to simultaneously segment multiple layers in severely diseased eyes.

In [42], a deep learning (DL) model that utilizes 3D convolutions to extract features from both spatial and inter-frame dimensions, enabling interpretation of data encoded in adjacent frames, is introduced for segmenting retinal layers in OCT scans automatically. By leveraging correlation information between neighboring frames, it generates predictions for the entire OCT volume in a single pass.

Hassan *et al* [43] put forward a combination of CNN and structure tensor-based segmentation frameworks (CNN-STSF) to segment eight retinal layers. The retinal layer information provided by the structure tensor enables the CNN model to focus on the pixels that are categorized as part of retinal layers rather than considering the entire pixels. The method utilized transfer learning where an AlexNet model [47] is used and is trained using image patches. In [44], the authors introduced a deep learning approach that effectively captures local and contextual information to segment the tissues of the optic nerve head. The method combined the skip connections from U-Net, residual learning and dilated convolution. The standard convolution blocks and residual blocks were built using dilated convolution layers. A segmented probability map is obtained after performing dilated convolution and upsampling.

In [45], a method that combines a fully convolutional network (FCN) with Gaussian processes for post-processing is proposed. The initial per-pixel estimates are obtained using a DenseNet-FCN with 103 layers. These estimates are then refined using Gaussian processes regression, incorporating a Radial Basis Function (RBF) kernel to model the smoothness of retinal surfaces. Furthermore, several multiscale CNN-based approaches [48, 49], which extract features by employing different kernel sizes at different scales and resolutions, have been suggested.

The performance of sequential data processing architectures like RNN [50] and LSTM [51] are also investigated in the retinal layer segmentation. In [50] a RNN is trained using image patches as a classifier along with a graph search algorithm to enhance the layer boundaries. A combination of CNN and LSTM is used in [51] for retinal layer segmentation. The features of images are extracted using CNN and LSTM is used for detecting the boundaries of layers.

In the realm of retinal layer segmentation, GAN-based architectures [52–54] have been explored. SGNet [52] employs an adversarial learning architecture for segmenting nine retinal layers and fluid regions in OCT B-scans in a semi-supervised manner. It utilizes unlabeled data and incorporates U-Net-like fully convolutional architecture in both the segmentation and discriminator networks. In [53], the authors investigated the effect of noise on semantic segmentation and employed a CycleGAN architecture to perform image-to-image translation and denoising. Furthermore, a GAN-based segmentation model is introduced for upscaling low-resolution OCT images to high resolution, combining ResNet and U-Net in the GAN architecture with additional sub-pixel convolution and transposed convolution layers [54]. In order to further improve the effectiveness of joint optimization jobs in the GAN, an additional reconstruction loss term called Dice loss has been included.

In [55], U-Net and advanced U-Net models such as U-Net++, VGG16-U-Net, VGG16-U-Net++, ResNet50-U-Net and ResNet50-U-Net++ are investigated for segmenting the retinal layers. The VGG16-U-Net network, which combines VGG16 and U-Net, outperformed the original U-Net and U-Net++ architectures. The models were trained on a cloud platform and a domain decomposition strategy used to reduce the complexity for processing large volumes of data in the cloud.

This chapter discusses in-depth the deep learning design cycle for retinal layer segmentation. Moreover, we employ the Python implementation of the U-Net [38] model for illustrative purposes.

5.3 Retinal layer segmentation using U-Net model

U-Net [38] has gained wide popularity in the field of segmentation of images. It includes a symmetric encoder and decoder. The expanding encoder extracts features from the input image and the contracting decoder reconstructs the segmentation map. Two 3×3 unpadded convolutions are applied in the encoder, and for downsampling, a rectified linear unit (ReLU) and a maximum pooling operation of 2×2 with stride 2 are applied after each convolution block. After each down-sampling step the number of feature channels are doubled. The decoder performs a 2×2 upsampling of feature maps which halves the number of feature channels. After performing two 3×3 convolutions and a ReLU, the result is concatenated with the matching feature maps from the encoder component. At the last layer in the decoder path, the 64-component feature vector is converted to necessary classes using a 1×1 convolution. The entire architecture includes 23 convolutional layers. Skip connections in the U-Net connect every pair of encoder-decoder layers, which enhances the gradient flow and improves data transfer between the up-sampling and down-sampling pathways.

An OCT image is used as the input for the U-Net model, which splits the image into multiple regions by assigning each pixel to one of several classes. The different classes are the various tissue layers and other areas near the tissue layers of interest. There are numerous literary works that draw inspiration from U-Net architecture, including [38–41, 47]. The original (vanilla) U-Net architecture has been modified in a number of ways to improve performance, including Dense U-Net [56], Inception U-Net [57], Attention U-Net [58], residual U-Net [59], recurrent-residual (R2) U-Net [60], U-Net++ [61], squeeze +excite (SE) U-Net [62], which are all investigated in [63] for retinal layer segmentation.

The U-Net architecture, used for retinal layer segmentation, follows a similar structure to the original U-Net with an encoder path and a decoder path. In the encoder path, a sequence of convolutions, rectified linear units (ReLU), and max pooling operations perform downsampling and extract feature maps. These feature maps are then upsampled in the decoder path using up-sampling layers and convolutions. To facilitate the utilization of fine-grained information from the encoder phase, skip connections are established between corresponding layers in the encoder and decoder parts. These connections allow for the fusion of detailed information in the decoder part. The final output is a probability map, where each pixel represents the likelihood of belonging to a specific class. Notably, U-Net differentiates from FCN by employing skip connections between each encoder-decoder level and instead of adding, the feature maps are concatenated.

5.4 Design cycle

The DL design cycle is an iterative process that includes experimentation, learning, and enhancement of models until the intended performance and outcomes are achieved. Generally, a DL design cycle constitutes problem definition, data collection and preprocessing, model definition, training and testing of the model, and performance evaluation.

The details on each of these steps are provided below.

5.4.1 Problem definition

Retinal layer segmentation is the method of identifying retinal layers from different retinal imaging modalities. Many retinal abnormalities can be diagnosed by analyzing the layers. Thus developing an automated retinal layer segmentation will help diagnosing the diseases. Here, we focus on the U-Net architecture for performing retinal layer segmentation using OCT images.

5.4.2 Data collection and preprocessing

A summary of various datasets for segmenting retinal layers are available in [37]. From the literature it is understood that Duke DME [28], a publicly available dataset, is used in a large number of works. In this study training and evaluation of the model is also conducted on the Duke DME OCT data set. The dataset comprises 10 patients' OCT scans that have been annotated by two clinicians. Each subject's image has 61 B-scans of size 496×768, out of which only 11 B-scans per patient are annotated for the retinal layers and fluid. Thus from 10 subjects, 110 Bscans are annotated. The resolution of each image is 496×768 and out of 110 images, 53 B-scans are of the pathological retina and 57 B-scans are of the normal retina. The Duke DME cyst dataset can be downloaded as a zip file from the url 'http://www.duke.edu/sf59/ Datasets/2015 BOE Chiu2.zip'. The dataset can be accessed using the **wget** command and unzip it using the **unzip** command. The corresponding code is given below.

```
wget http://www.duke.edu/~sf59/Datasets/2015_BOE_Chiu2.zip
unzip 2015_BOE_Chiu2.zip
```

The code can be written as a shell script(.sh) or it can be directly executed in Jupyter Notebook by prefixing an exclamation mark before the commands.

The OCT images of 10 subjects are provided in .mat format (matlab file) in the Duke dataset. Each subject's file is organized as a dictionary with keys 'images', 'automaticFluidDME', 'manualFluid1', 'manualFluid2', 'automaticLayersDME', 'automaticLayersNormal' and 'manualLayers2'. 'images' provides OCT images with dimension [61, 496, 768]. The annotations of retinal layers and fluid are performed by two experts and are available in 'manualLayers1', 'manualLayers2', 'manualFluid1' and 'manualFluid2'. The automatic annotations are also available in 'automaticLayersDME', 'automaticLayersNormal' and 'automaticFluidDME'.

The annotations with keys 'automaticLayersDME', 'automaticLayersNormal ','manualLayers1', and 'manualLayers2', are with dimensions ([8, 61, 768]) in which the first dimension 8 describes the annotations in the form of range of pixel values for each layer. These annotations can be used for creating masks for each image, which will be used as the ground truth during training. The code snippet for displaying the annotation range is given below.

5.4.2.1 Displaying the sample annotation—Python code

```
#import necessary library
from scipy.io import loadmat
#load matlab file
octimage = loadmat('/2015_BOE_Chiu/Subject_01.mat')
# Load annotations of retinal layers by Expert 1
annotation=octimage['manualLayers1']
print(annotation[:,118,10])
```

Subject1's 10 th B-scan annotation values are starting from the 118 th column and the column values are shown as output. The output of the above code is

[185. 195. 204. 210. 217. 235. 243. 248].

The pixel value ranges can be used to identify specific regions within the retinal layers. For instance, pixel values from 0 to 184 correspond to the RaR and values from 185 to 194 represent the ILM. The ranges continue as follows: 195 to 203 for NFL-IPL layer, 204 to 209 for the INL, 210 to 216 for OPL, 217 to 234 for ONL-ISM, 235 to 242 for ISE, 243 to 247 for OS-RPE, and 248 onwards for the RbR. For each B-scan, a corresponding mask is generated, serving as a label for training. The OCT scans of each patient have dimensions (496, 768, 61), with only 11 B-scans being annotated. These 11 B-scans are extracted, and a mask is created for each image. The function **get_valid_img_mask** is performing this. The manual_l and manual_f variables store the annotations, and the **get_valid_index** function filters and returns only the B-scans with valid annotations.

5.4.2.2 Valid image and mask generation—Python code

```python
def get_valid_img_mask(scan):
    fluid_cls = 9
    manual_l = np.array(scan['manualLayers1'], dtype=np.uint16)
    manual_f = np.array(scan['manualFluid1'], dtype=np.uint16)
    image = np.array(scan['images'], dtype=np.uint8)
    annotated_idx = get_annotated_idx(manual_l)
    #get_annotated_idx function will extract the indices of 11 Bscans
    #which is having annotations
    manual_f = manual_f[:, :, annotated_idx]
    manual_l = manual_l[:, :, annotated_idx]
    mask = np.zeros_like(manual_f, dtype=np.uint8)
    # To avoid 0 added for empty class
    for bscan in range(mask.shape[2]):
        for ascan in range(mask.shape[1]):
            class_index = manual_l[:,ascan, bscan]
            for j,_ in enumerate(class_index):
                if j > 0 and class_index[j] < class_index[j-1]:
                    class_index[j] = class_index[j-1]
            # Mask generation-layers are given values from 1 to 7
            # 0 for Region Above Retina and 8 for Region Below RPE
            for label, (index_prev, index_cur) in enumerate(
                zip([0,*class_index], [*class_index, mask.shape[0]])):
                mask[index_prev:index_cur, ascan, bscan] = label
    # Fluid annotated pixels are assigned fluid class value 9
    mask[manual_f > 0] = fluid_cls
    # Getting the region where annotaion is provided
    a_scan_ann, = np.where(np.sum(manual_l, axis=(0,2)) != 0)
    mask = mask[:, a_scan_ann[0]:a_scan_ann[-1] + 1]
    image = image[:, a_scan_ann[0]:a_scan_ann[-1] + 1]
    # only return images with annotations
    image = image[:,:,annotated_idx]
    return image, mask
```

To handle the mistakenly added 0's for empty classes in the annotations, just the previous value of annotation is copied down to that position. As per the annotation, the range of pixel values in the mask are assigned with class labels—0 for Region above retina, 1–7 for 7 retinal layers, 8 for Region below retina and 9 for fluid. The annotation range varies among the patients and that is also handled at the end of the code.

For this study, we used the Duke DME dataset in two ways. First approach is to use the B-scan images for training, testing and validation. The Bscans (11 B-scans per patient) are generated using the function **bscan_slicing**. Here the patient's OCT scan with annotations is given as input and the output is a list of all B-scans. Second approach is to slice the OCT B-scans into patches of width 64 pixels (496 × 64) and use them for training, validation and testing. Slicing the images vertically into 64-width pixel patches helps to tackle the dataset's limited size and reduces the risk of overfitting. As a result, 858 images of size [496 × 64] are obtained from this process. The function "slicing" is performing this operation. The input to the slicing function is the OCT scan of a patient and output is the slices generated as a list.

5.4.2.3 Image slicing—Python code

```python
def bscan_slicing(scan)->list:
    images = []
    # Append each Bscan to image list
    for i in range(scan.shape[2]):
        images.append(scan[:,:,i])
    return images


def slicing(scan)->list:
    images = []
    # Bscans are appended to the list image
    for k in range(scan.shape[2]):
        images.append(scan[:,:,k])
    # Bscans are sliced to 496 X 64
    slice_width=64
    slices = []
    for bscan in images:
        for j in range(bscan.shape[1] // slice_width):
            slices.append(bscan[:,j*slice_width:(j+1)*slice_width])
    return slices
```

5.4.3 Model definition

The model implemented in this work is using PyTorch, which is an open source framework for developing DL networks. PyTorch represents multidimensional data as tensors, which has the capability of working on both GPU and CPU. PyTorch provides various classes that can be used for developing and training neural architectures. A basic neural network architecture can be built using **torch.nn. Module** class and corresponding **forward** function.

The U-Net model defined in our work also has four encoder and decoder blocks each. In this instance, the size of the input image and the number of channels used for the convolution blocks of the encoder and decoder blocks are different. The B-scans are resized to 224 × 224 and given as input similar to the method used in [64]. The number of channels used in the first, second, third and fourth encoders are 32, 64, 128 and 256, respectively. In the bottleneck layer there are 512 channels. During the decoding phase the number of channels are halved in the order 256, 128, 64 and 32. At the end of the last decoder, softmax function is applied and the segmentation map is created with 9 classes—7 retinal layers, fluid and background. The required modules can be imported initially using the statements:

```
import torch
import torch.nn as nn
```

The encoder block consists of two convolution layers with 3 × 3 kernels and padding 1, followed by batch normalization and ReLU activation. The convolutional layers in the encoder block can be expressed as follows:

Convolution block—Python code

```python
import torch
import torch.nn as nn
class conv_block(nn.Module):
    def __init__(self, input_channels,features):
        super().__init__()
        self.conv_1 = nn.Conv2d(input_channels,features,
                    kernel_size=3, padding=1, bias=False)
        self.bnorm_1 = nn.BatchNorm2d(features)
        self.relu_1 = nn.ReLU(inplace=True)
        self.conv_2 = nn.Conv2d(features, features,
                    kernel_size=3, padding=1,bias=False)
        self.bnorm_2 = nn.BatchNorm2d(features)
        self.relu_2 = nn.ReLU(inplace=True)
    def forward(self, input):
        output = self.conv_1(input)
        output = self.bnorm_1(output)
        output = self.relu_1(output)
        output = self.conv_2(output)
        output = self.bnorm_2(output)
        conv_blk = self.relu_2(output)
        return conv_blk
```

__init__() is a constructor method defined inside any class, which is used to initialize all the parameters inside the class. This method is invoked when an object of the class is created. **super()** is a function defined inside the __init__() function, which is used to call the __init__() function of the parent class, here the parent class is **nn.Module.** Convolutional layers, Batch normalization and ReLu can be defined using predefined functions inside the `torch.nn` such as `nn.Conv2d`, `nn.BatchNorm2d` and `nn.ReLU`. The **forward** () function represents the forward pass of the network and it returns an object for the model created.

Encoder block—Python code

Here, the encoder blocks can be created as per the code given. After each encoder block, there is a Maxpool layer with stride 2 and kernel size 2.

```
class encoder_block(nn.Module):
    def __init__(self, input_channels,features):
        super(encoder_block,self).__init__()
        self.conv_1 = conv_block(input_channels,features)
        self.pool_1 = nn.MaxPool2d(kernel_size=2, stride=2)
    def forward(self, inputs):
        conv_out = self.conv_1(inputs)
        mp_out = self.pool_1(conv_out)
        return conv_1_out, mp_out
```

The decoder block consists of two convolution layers with 3×3 kernels and padding 1, similar to the encoder block. Upsampling is achieved through a transposed convolution with a kernel size of 2 and a stride of 2. The decoder block can be defined as follows.

Decoder block—Python code

```
class decoder_block(nn.Module):
    def __init__(self, input_channels, features):
        super(decoder_block,self).__init__()
        self.up_conv_1 = nn.ConvTranspose2d(input_channels,
                        features, kernel_size=2, stride=2)
        self.conv_1 = conv_block((features * 2), features)
    def forward(self, inputs, skip_con):
        output = self.up_conv_1(inputs)
        output = torch.cat((output, skip_con), axis=1)
        output = self.conv_1(out)
        return output
```

Following the last decoder, a convolution layer with kernel size 1 and a softmax function is applied. The encoder decoder blocks are connected together at bottom by a bottleneck layer. The whole network can be coded as follows.

U-Net Network—Python code

```python
class Unet(nn.Module):
    def __init__(self, input_channels=1, output_channels=1,
                                            features=32):
        super(Unet,self).__init__()
        # Encoder Blocks
        self.enc1 = encoder_block(input_channels, features)
        self.enc2 = encoder_block(features, features * 2)
        self.enc3 = encoder_block(features * 2, features * 4)
        self.enc4 = encoder_block(features * 4, features * 8)
        # Bottleneck layer
        self.bottl = conv_block(features * 8, features * 16)
        #  Decoder blocks
        self.dec1 = decoder_block( features * 16, features * 8)
        self.dec2 = decoder_block( features * 8, features * 4)
        self.dec3 = decoder_block( features * 4, features * 2)
        self.dec4 = decoder_block( features * 2, features)
        # output layer
        self.outputs = nn.Conv2d(features, output_channels,
                                    kernel_size=1, padding=0)
        self.softmax = nn.Softmax2d()
    def forward(self, inputs):
        enc1_out, pool1 = self.enc1(inputs)
        enc2_out, pool2 = self.enc2(pool1)
        enc3_out, pool3 = self.enc3(pool2)
        enc4_out, pool4 = self.enc4(pool3)
        bottle_neck = self.bottl(pool4)
        dec1_out = self.dec1(bottle_neck, enc4_out)
        dec2_out = self.dec2(dec1_out, enc3_out)
        dec3_out = self.dec3(dec2_out, enc2_out)
        dec4_out = self.dec4(dec3_out, enc1_out)
        output = self.outputs(dec4_out)
        output = self.softmax(output)
        return output
```

5.4.4 Model training

The U-Net Model can be generated by the following code, where argument 1 is the number of input channels.

$$\text{model} = \text{U} - \text{Net} \qquad (5.1)$$

Once the model is created, the training needs a dataset with uniform size, because the annotations for the various patients are of different sizes. Each of the 110 (11×10) images are scaled to [224×224] after performing normalization. Training starts by the statement **model.train()**. The code for training is given in the function **train(args)**. The arguments to the function can be implemented using an **argparse** library. The argument values used in our experiments are **device** = 'cuda' or 'cpu', **n_classes** = 9, **data_path** = path to dataset, **iterations**=100 (epoch), **image_size**=224 and **batch_-size** = 10. Adam optimizer is used with learning rate $5e^{-4}$ and a weight decay of $1e^{-4}$. Function **get_data()** will perform the required transformations such as resizing of images and masks to 224 × 224 and normalization. **DataLoader** is a built-in class that provides an efficient way to load the data into the model. Prediction is performed by using the statement **pred = model (image)**. The whole training process can be coded as follows.

U-Net network training—Python code

```python
def train(args):
    device = args.device
    n_classes = args.n_classes
    learning_rate = args.learning_rate
    data_path = args.data_path
    iterations = args.num_iterations
    img_size = args.image_size
    batch_size = args.batch_size
    criterion_seg = CombinedLoss()
    trainingEpoch_loss=[]
    validationEpoch_loss=[]
    save_name = "Unet" + ".pt"
    max_dice_score = 0
    best_test_dice = 0
    best_iter = 0
    model = get_model("unet",num_classes=n_classes).to(device)
    model.train()
    optimizer = torch.optim.Adam(list(model.parameters()),
            lr=learning_rate, weight_decay=args.weight_decay)
    train_data_loader, val_data_loader, test_data_loader, _, _, _ = \
                get_data(data_path, img_size, batch_size)
```

```python
for t in range(iterations):
    step_loss = []
    for image, label in tqdm.tqdm(train_data_loader):
        image = image.to(device)
        label = label.to(device)
        pred = model(image)
        loss = criterion_seg(pred, label.squeeze(1), device=device)
        loss.backward()
        optimizer.step()
        step_loss.append(loss.item())

    print("Epoch No. ", t, "/", iterations)
    print("Validation Testing")
    dice_score,vloss = evaluation(val_data_loader, criterion_seg,
                          model, n_classes=n_classes, device=device)
    print("Testing with Expert 1 annotation")
    dice_test_score = evaluation(test_data_loader, criterion_seg,
                          model, n_classes=n_classes, device=device)
    if dice_score > max_dice_score:
        max_dice_score = dice_score
        best_iter = t
        best_test_dice = dice_test_score
        print("Updating model, epoch: ", t)
        torch.save(model.state_dict(), save_name)
    model.train()
    trainingEpoch_loss.append(np.array(step_loss).mean())
    validationEpoch_loss.append(vloss)
print("Best iteration: ", best_iter, "Best val dice: ", max_dice,
                          "Best test dice: ", best_test_dice)

epochs = range(1, t+2)
# Plot Training and validation Loss curves
plt.plot(epochs, trainingEpoch_loss, label='Training Loss')
plt.plot(epochs, validationEpoch_loss, label='Validation Loss')
# Add in a title and axes labels
plt.title('Training and Validation Loss')
plt.xlabel('Epochs')
plt.ylabel('Loss')
# Display the plot
plt.legend(loc='best')
plt.savefig('./loss_plot/lossvsepoch.png',bbox_inches="tight")
return model
```

During training, after each iteration (epoch) validation testing is performed and corresponding validation loss and Dice score are calculated. The code also performs testing using annotations by expert 1 on every epoch with maximum Dice score and that model is saved. The model can be saved for future use by the statement **torch.save(model.state_dict(), save_name)**, where **save_name** is the saved name of the model. Validation loss and training loss during each epoch is stored and the loss curve is plotted at the end.

The loss function is a combination of Dice loss and Cross entropy loss. The Cross entropy loss is calculated using the function **CrossEntropyLoss()** defined in the **torch. nn** module. The arguments are the pred and target, where pred is the prediction done by the model and target is the actual map.

Loss calculation—Python code

```
class CrossEntropyLossCal(nn.Module):
    def __init__(self, weight=None, size_average=True):
        super(CrossEntropyLossCal, self).__init__()
        self.celoss = nn.CrossEntropyLoss(weight, size_average)
    def forward(self, pred, target):
        return self.celoss(pred, target)
class CombinedLoss(nn.Module):
    def __init__(self):
        super(CombinedLoss, self).__init__()
        self.cross_entropy_1= CrossEntropyLossCal()
        self.dice_loss = DiceLoss()
    def forward(self, pred, target,weight=1, device="cuda"):
        target = target.type(torch.LongTensor).to(device)
        pred_soft = fun.softmax(pred,dim=1)
        l2 = torch.mean(self.dice_loss(pred_soft, target))
        l1 = torch.mean(torch.mul(self.cross_entropy_1.forward(pred, target),weight))
        l = l1 + l2
        return l
```

The dice loss calculation is done as follows:

```
class DiceLoss(_Loss):
    def forward(self, output, target):
        epsilon = 0.0001
        output = output.exp()
        encd_target = output.detach() * 0
        encd_target.scatter_(1, target.unsqueeze(1), 1)
        intersection= output * encd_target
        numer = 2 * intersection.sum(0).sum(1).sum(1)
        denom = output + encd_target
        denom = denom.sum(0).sum(1).sum(1) + epsilon
        loss_per_channel = (1 - (numer / denom))
        return loss_per_channel.sum() / output.size(1)
```

Evaluation of the model—Python code

```python
def evaluation(eval_data_loader, criterion, model, n_classes, device):
    model.eval()
    loss = 0
    counter = 0
    dice = 0
    dice_all = np.zeros(n_classes)
    for image, label in tqdm.tqdm(eval_data_loader):
        image = image.to(device)
        label = label.to(device)
        label_oh = torch.nn.functional.one_hot(label,
                                               num_classes=n_classes)
        pred = model(image)
        max_val, idx = torch.max(pred, 1)
        pred_oh = torch.nn.functional.one_hot(idx, num_classes=n_classes)
        dice1, dice2 = dice_each_class(pred_oh, label_oh, n_classes)
        dice += dice1
        dice_all += dice2
        loss += criterion(pred, label.squeeze(1), device=device).item()
        counter += 1
    loss = loss / counter
    dice = dice / counter
    dice_all = dice_all / counter
    print("Validation loss: ", loss, " Mean Dice: ", dice.item(),
                              "Dice All:", dice_all)
    return dice,loss
```

The **evaluation(eval_data_loader, criterion, model, n_classes, device)** function will evaluate the model by calculating the Dice score. The argument **eval_data_-loader** is the set of data to be evaluated. **criterion** is a function that returns the loss, **n_classes** is the number of classes, which is 9, and the device can be either 'cuda' or 'cpu'. The overall Dice score, Dice score for individual classes and loss are calculated for each batch and after processing all batches, average loss and Dice values are calculated.

Dice score calculation is done by the function **dice_each_class(pred, true, n_class)**. The argument **pred** is the one hot tensor of the prediction done by the model, **true** is the one hot tensor corresponding to the true label and **n_class** is the number of classes. The return values are the average Dice score (**average_dice**) and the Dice score corresponding to each class as a list (**dice_all**).

```
def dice_each_class(pred, true, n_class):
    average_dice = 0
    pred = pred.data.squeeze()
    true = true.data.squeeze()
    dice_all = np.zeros(n_class)
    for i in range(n_class):
        True = true[:,:,i].view(-1)
        Pred = pred[:,:,i].view(-1)
        inter = (True * Pred).sum() + 0.0001
        union = True.sum() + Pred.sum() + 0.0001
        T = 2 * inter / union
        average_dice = average_dice + (T / n_class)
        dice_all[i] = T
    return average_dice, dice_all
```

5.4.5 Experimental results

As discussed earlier, the U-Net model in our work is trained with the Duke DME dataset in two ways—using B-scans and by using B-scan slices of size 496 ×64—for training, validation and testing. The models obtained after both the approaches are evaluated solely for explanatory purposes.

5.4.5.1 Experimental setup
Similar to the approaches used in [64], training is performed using scans from the first six patients. The scans of patient 7 and 8 are used for validation and remaining scans from patient 9 and 10 for testing. The models are evaluated using the annotations provided by expert 1 in the Duke dataset. To evaluate the model, loss curves are plotted, and both qualitative and quantitative analyses are conducted.

A comparative study of UNet models with different filter sizes is also performed. The number of channels/filters, F, in the first encoder block of the UNet model is experimented with different filter sizes—16, 32 and 64 and subsequently the number of filters in the following encoder and decoder blocks of the UNet are modified accordingly. For example, if the number of filters, F in the first encoder block is 16, then there are 32, 64, 128, 256, 128, 64, 32 and 16 filters in the encoder2, encoder3, encoder4, bottleneck, decoder1, decoder2, decoder3 and decoder4 blocks, respectively. Thus the depth of the input is increasing up to the bottleneck layer and then the depth is decreasing until we get the final output. Same approach is followed for $F = 32$ and $F= 64$.

5.4.5.2 Learning curves

Generally, learning curves provide a measure of the progress of learning over time. In any supervised learning method, for improving the model, parameter weights and biases are updated while trying to minimize the loss function. Thus if we plot the loss curve during training, it depicts how well our model is learning. Loss curves generally plot loss functions on Y-axis and number of epochs on X-axis. During the initial stages of plot a sharp decline indicates the model is improving its performance. When the number of epochs increases, the rate of improvement slows down gradually. Thus the training loss curve demonstrates how well the model is working on training data. To evaluate the performance of the model on unseen data, a validation testing can be performed, whose result is used for improving the design or for performing hyper parameter tuning. The plotting of validation loss along with training loss helps to identify issues with learning such as underfit or overfit. The train and validation loss curves for our models are given in figure 5.2. From the curves we can see that slicing the B-scan is improving the convergence speed.

The training and validation loss curves with different filter sizes are shown in figure 5.3. Figure 5.3(a) shows the loss curve plot of the model using $F = 16$. Figure 5.3(b) gives the plot for 32 filters and figure 5.3(c) shows the plot while using 64 filters. From the figures we can see that when $F = 16$ the model is taking more time to converge and when $F = 32$ and $F = 64$ models are performing much better. But when $F = 64$ the loss plot has lots of fluctuations, which can be the result of noisy data or overfitting.

5.4.5.3 Quantitative analysis

A quantitative analysis will provide a numerical measure for the performance of the model. Dice score is used as the quantitative evaluation method for the model in our experiments. The Dice score is a popular performance metric for applications involving segmentation. It measures the similarity between the expected segmentation mask and the actual segmentation mask. The Dice score goes from 0 to 1, where a value of 1 denotes a complete resemblance between the expected and actual results. The Dice score can be formulated as follows.

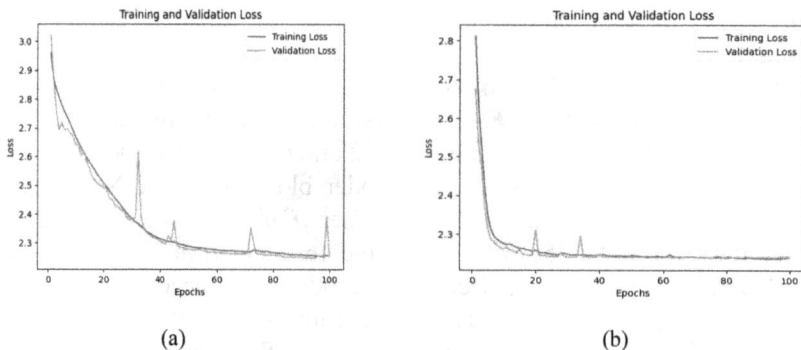

(a)   (b)

Figure 5.2. Loss curves for the U-Net model. (a) Trained with Bscans (b) Trained with slices of size (496 ×64).

(a)

(b)

(c)

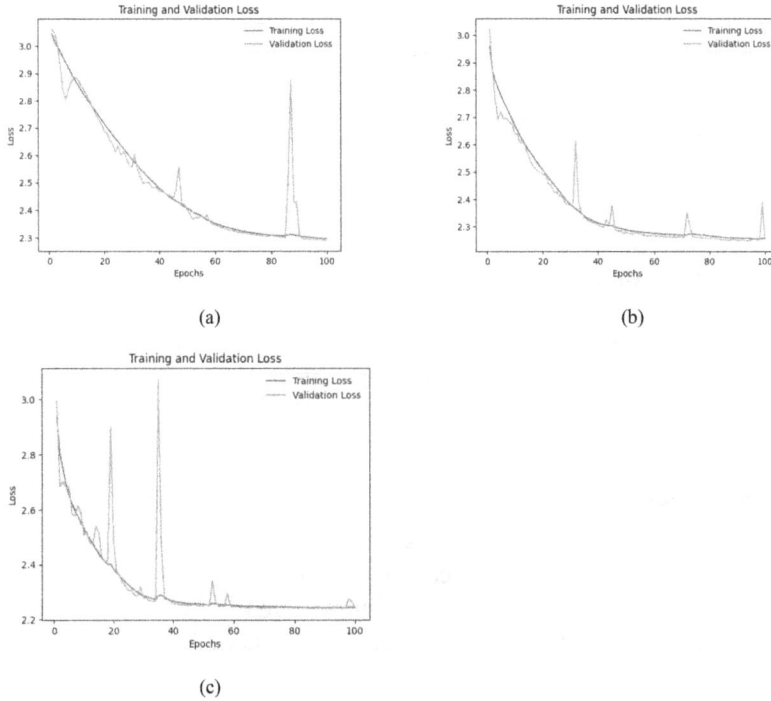

Figure 5.3. Training and validation loss curves for the U-Net model with different filter sizes. (a) $F = 16$ (b) $F = 32$ (c) $F = 64$.

Table 5.1. Quantitative analysis using Dice score.

	Dice scores for different models			
	$F = 16$	$F = 32$	$F = 64$	$F = 32$ and Bscan slice with size (496 × 64)
Validation Dice score	0.60	0.899	0.91	0.955
Test Dice score	0.54	0.835	0.84	0.908

$$\text{Dice score} = \frac{2 \times \text{intersection}}{\text{union} + \text{intersection}} \qquad (5.2)$$

The term *intersection* represents the pixels present in both the result and ground truth. *Union* indicates the pixels present either in the result or the ground truth.

The models are evaluated using Dice score values. The Dice score values obtained for the model using different approaches are given in table 5.1. The best validation and test Dice score for UNet models with different filter sizes that are trained with Bscans are given in the first three columns. The Dice scores for the model with $F = 32$ and trained using slices of size 496 × 64 are given in the last column of the table.

From the table it is clear that the model trained with slices of 64 pixel width and having $F = 32$ is giving a better performance.

5.5 Conclusion

OCT images are essential for early diagnosis of various retinal and neurodegenerative disorders. Segmentation of OCT images, especially retinal layer segmentation is vital, in this diagnosis. Compared to traditional methods, deep learning methods or the approaches combining traditional and DL methods are giving better results. An overview of the research work carried out in this area and a simple implementation of retinal layer segmentation using U-Net architecture is presented in this work. OCT image segmentation using deep learning approaches has a promising future.

Acknowledgments

The authors would like to express their sincere gratitude to the authors and creators of the online sources that greatly contributed to the development of this book chapter. Also, the authors would like to acknowledge the scholarly articles, research papers, and websites that provided invaluable insights and information.

References

[1] Zhu J, Zhang E and Rio-Tsonis K D 2012 *Eye Anatomy* (New York: Wiley Online)

[2] Li Y, Xia X and Paulus Y M 2018 Advances in retinal optical imaging *Photonics* **5** 9

[3] Hertzmark E *et al* 2003 *Optical coherence tomography measurement of macular and nerve fiber layer thickness in normal and glaucomatous human eyes Ophthalmology* **110** 177–89

[4] Jivrajka R V, Chang K T, Alasil T, Walsh A C, Sadda S R, Keane P A and Liakopoulos S 2009 *Evaluation of optical coherence tomography retinal thickness parameters for use in clinical trials for neovascular age-related macular degeneration Invest. Ophthalmol. Vis. Sci.* **50** 3378–85

[5] Clay Bavinger J, Dunbar G E, Stem M S, Blachley T S, Kwark L, Farsiu S, Jackson G R and Gardner T W 2016 *The effects of diabetic retinopathy and pan-retinal photocoagulation on photoreceptor cell function as assessed by dark adaptometry Invest. Ophthalmol. Vis. Sci.* **57** 208–17

[6] Abràmoff M D, Garvin M K and Sonka M 2010 *Retinal imaging and image analysis IEEE Rev. Biomed. Eng.* **3** 169–208

[7] Bhende M, Shetty S, Kuppuswamy Parthasarathy M and Ramya S 2018 *Optical coherence tomography: a guide to interpretation of common macular diseases Indian J. Ophthalmol.* **66** 20

[8] Patel P J, Chen F K, da Cruz L and Tufail A 2009 *Segmentation error in stratus optical coherence tomography for neovascular age-related macular degeneration Invest. Ophthalmol. Vis. Sci.* **50** 399–404

[9] Fu T, Liu X, Liu D and Yang Z 2017 *A deep convolutional feature based learning layer-specific edges method for segmenting oct image 9th Int. Conf. on Digital Image Processing (ICDIP 2017)* **vol 10420** (Bellingham, WA: SPIE) pp 480–4

[10] Mohandass G, Natarajan R A and Sendilvelan S 2017 *Retinal layer segmentation in pathological sd-oct images using boisterous obscure ratio approach and its limitation Biomed. Pharm. J.* **10** 1585–91

[11] Swanson E A and Fujimoto J G 2017 *The ecosystem that powered the translation of oct from fundamental research to clinical and commercial impact Biomed. Opt. Express* **8** 1638–64

[12] Shahidi M, Wang Z and Zelkha R 2005 *Quantitative thickness measurement of retinal layers imaged by optical coherence tomography Am. J. Ophthalmol.* **139** 1056–61

[13] Tan O, Li G, Lu A T-H, Varma RDavid Huang Advanced Imaging for Glaucoma Study Group *et al* 2008 *Mapping of macular substructures with optical coherence tomography for glaucoma diagnosis Ophthalmology* **115** 949–56

[14] de Sisternes L, Jonna G, Moss J, Marmor M F, Leng T and Rubin D L 2017 *Automated intraretinal segmentation of sd-oct images in normal and age-related macular degeneration eyes Biomed. Opt. Express* **8** 1926–49

[15] Yin X, Chao J R and Wang R K 2014 *User-guided segmentation for volumetric retinal optical coherence tomography images J. Biomed. Opt.* **19** 086020 086020

[16] Sun Y, Niu S, Gao X, Su J, Dong J, Chen Y and Wang L 2020 *Adaptive-guided-coupling-probability level set for retinal layer segmentation IEEE J. Biomed. Health Inform.* **24** 3236–47

[17] Kajic V, Povazay B, Hermann B, Hofer B, Marshall D, Rosin P L and Drexler W 2010 *Robust segmentation of intraretinal layers in the normal human fovea using a novel statistical model based on texture and shape analysis Opt. Express* **18** 14730–44

[18] Lang A, Carass A, Hauser M, Sotirchos E S, Calabresi P A, Ying H S and Prince J L 2013 *Retinal layer segmentation of macular oct images using boundary classification Biomed. Opt. Express* **4** 1133–52

[19] Mayer M A, Tornow R P, Hornegger J and Kruse F E 2008 *Fuzzy c-means clustering for retinal layer segmentation on high resolution oct images BIOSIGNAL 2008 : 19th Biennial Int. EURASIP Conf.* (Eur: EURASIP)

[20] Fang L, Cunefare D, Wang C, Guymer R H, Li S and Farsiu S 2017 *Automatic segmentation of nine retinal layer boundaries in oct images of non-exudative amd patients using deep learning and graph search Biomed. Opt. Express* **8** 2732–44

[21] Shah A, Zhou L, Abrámoff M D and Wu X 2018 *Multiple surface segmentation using convolution neural nets: application to retinal layer segmentation in oct images Biomed. Opt. Express* **9** 4509–26

[22] Ishikawa H, Stein D M, Wollstein G, Beaton S, Fujimoto J G and Schuman J S 2005 *Macular segmentation with optical coherence tomography Invest. Ophthalmol. Vis. Sci.* **46** 2012–7

[23] Cabrera Fernández D, Salinas H M and Puliafito C A 2005 *Automated detection of retinal layer structures on optical coherence tomography images Opt. Express* **13** 10200–16

[24] Yazdanpanah A, Hamarneh G, Smith B and Sarunic M 2009 *Intra-retinal layer segmentation in optical coherence tomography using an active contour approach Medical Image Computing and Computer-Assisted Intervention–MICCAI 2009: 12th Int. Conf.* (Berlin: Springer) 649–56

[25] Ghorbel I, Rossant F, Bloch I, Tick S and Paques M 2011 *Automated segmentation of macular layers in oct images and quantitative evaluation of performances Pattern Recognit.* **44** 1590–603

[26] Yang Q, Reisman C A, Wang Z, Fukuma Y, Hangai M, Yoshimura N, Tomidokoro A, Araie M, Raza A S, Hood D C *et al* 2010 *Automated layer segmentation of macular oct images using dual-scale gradient information Opt. Express* **18** 21293–307

[27] Chiu S J, Li X T, Nicholas P, Toth C A, Izatt J A and Farsiu S 2010 *Automatic segmentation of seven retinal layers in sdoct images congruent with expert manual segmentation Opt. Express* **18** 19413–28

[28] Chiu S J, Allingham M J, Mettu P S, Cousins S W, Izatt J A and Farsiu S 2015 *Kernel regression based segmentation of optical coherence tomography images with diabetic macular edema Biomed. Opt. Express* **6** 1172–94

[29] Stephanie J, Chiu J A, Izatt R V, O'Connell K P, Winter C A, Toth and Farsiu S 2012 *Validated automatic segmentation of amd pathology including drusen and geographic atrophy in sd-oct images Invest. Ophthalmol. Vis. Sci.* **53** 53–61

[30] Dufour P A, Ceklic L, Abdillahi H, Schroder S, De Dzanet S, Wolf-Schnurrbusch U and Kowal J 2012 *Graph-based multi-surface segmentation of OCT data using trained hard and soft constraints IEEE Trans. Med. Imaging* **32** 531–43

[31] Goutham A, Reddy B G, Kosuri R, Praneeth P and Lavanya R 2022 *A novel approach to segment oct layers using graph method for glaucoma diagnosis 2022 6th Int. Conf. on Trends in Electronics and Informatics (ICOEI)* (Piscataway, NJ: IEEE) pp 1588–90

[32] Vermeer K A, Van der Schoot J, Lemij H G and de Boer J F 2011 *Automated segmentation by pixel classification of retinal layers in ophthalmic oct images Biomed. Opt. Express* **2** 1743–56

[33] Srinivasan P P, Heflin S J, Izatt J A, Arshavsky V Y and Farsiu S 2014 *Automatic segmentation of up to ten layer boundaries in sd-oct images of the mouse retina with and without missing layers due to pathology Biomed. Opt. Express* **5** 348–65

[34] Karri S P K, Chakraborthi D and Chatterjee J 2016 *Learning layer-specific edges for segmenting retinal layers with large deformations Biomed. Opt. Express* **7** 2888–901

[35] Xiang D, Chen G, Shi F, Zhu W, Liu Q, Yuan S and Chen X 2018 *Automatic retinal layer segmentation of oct images with central serous retinopathy IEEE J. Biomed. Health Inform.* **23** 283–95

[36] Liu X, Fu T, Pan Z, Liu D, Hu W, Liu J and Zhang K 2018 *Automated layer segmentation of retinal optical coherence tomography images using a deep feature enhanced structured random forests classifier IEEE J. Biomed. Health Inform.* **23** 1404–16

[37] Viedma I A, Alonso-Caneiro D, Read S A and Collins M J 2022 *Deep learning in retinal optical coherence tomography (OCT): a comprehensive survey Neurocomputing* **507** 247–64

[38] Ronneberger O, Fischer P and Brox T 2015 *U-Net: convolutional networks for biomedical image segmentation Medical Image Computing and Computer-Assisted Intervention–MICCAI 2015: 18th Int. Conf., Munich* (Berlin: Springer) 234–41

[39] Roy A G, Conjeti S, Karri S P K, Sheet D, Katouzian A, Wachinger C and Navab N 2017 *Relaynet: retinal layer and fluid segmentation of macular optical coherence tomography using fully convolutional networks Biomed. Opt. Express* **8** 3627–42

[40] He Y, Carass A, Yun Y, Zhao C, Jedynak B M, Solomon S D, Saidha S, Calabresi P A and Prince J L 2017 *Towards topological correct segmentation of macular oct from cascaded FCNS Fetal, Infant and Ophthalmic Medical Image Analysis: Int. Workshop, FIFI 2017, and 4th Int. Workshop, OMIA 2017, Held in Conjunction with MICCAI 2017* (Berlin: Springer) 202–9

[41] Apostolopoulos S, De Zanet S, Ciller C, Wolf S and Sznitman R 2017 *Pathological oct retinal layer segmentation using branch residual u-shape networks Medical Image Computing and Computer Assisted Intervention—MICCAI 2017: 20th Int. Conf., Quebec City, QC, Canada* (Berlin: Springer) 294–301

[42] Kiaee F, Fahimi H and Rabbani H 2018 *Intra-retinal layer segmentation of optical coherence tomography using 3D fully convolutional networks 2018 25th IEEE Int. Conf. on Image Processing (ICIP)* (Piscataway, NJ: IEEE) 2795–9

[43] Hassan T, Usman A, Usman Akram M, Furqan Masood M and Yasin U 2018 *Deep learning based automated extraction of intra-retinal layers for analyzing retinal abnormalities 2018 IEEE 20th Int. Conf. on e-Health Networking, Applications and Services (Healthcom)* (Piscataway, NJ: IEEE) 1–5

[44] Krishna Devalla S *et al* 2018 DrU-Net: a dilated-residual U-Net deep learning network to digitally stain optic nerve head tissues in optical coherence tomography images arXiv:1803.00232

[45] Pekala M, Joshi N, Alvin Liu T Y, Bressler N M, DeBuc D C and Burlina P 2019 *Deep learning based retinal oct segmentation Comput. Biol. Med.* **114** 103445

[46] Long J, Shelhamer E and Darrell T 2015 *Fully convolutional networks for semantic segmentation Proc. IEEE Conference on Computer Vision and Pattern Recognition* (Piscataway, NJ: IEEE) 3431–40

[47] Krizhevsky A, Sutskever I and Hinton G E 2017 *Imagenet classification with deep convolutional neural networks Commun. ACM* **60** 84–90

[48] Hu K, Shen B, Zhang Y, Cao C, Xiao F and Gao X 2019 *Automatic segmentation of retinal layer boundaries in oct images using multiscale convolutional neural network and graph search Neurocomputing* **365** 302–13

[49] Chen M, Ma W, Shi L, Li M, Wang C and Zheng G 2021 *Multiscale dual attention mechanism for fluid segmentation of optical coherence tomography images Appl. Opt.* **60** 6761–8

[50] Kugelman J, Alonso-Caneiro D, Read S A, Vincent S J and Collins M J 2018 *Automatic segmentation of oct retinal boundaries using recurrent neural networks and graph search Biomed. Opt. Express* **9** 5759–77

[51] Gopinath K, Rangrej S B and Sivaswamy J 2017 *A deep learning framework for segmentation of retinal layers from oct images 2017 4th IAPR Asian Conf. on Pattern Recognition (ACPR)* pp 888–93

[52] Liu X, Cao J, Fu T, Pan Z, Hu W, Zhang K and Liu J 2019 *Semi-supervised automatic segmentation of layer and fluid region in retinal optical coherence tomography images using adversarial learning IEEE Access* **7** 3046–61

[53] Viedma I A, Alonso-Caneiro D, Read S A and Collins M J 2021 *Oct retinal image-to-image translation: analysing the use of cyclegan to improve retinal boundary semantic segmentation 2021 Digital Image Computing: Techniques and Applications (DICTA)* (Piscataway, NJ: IEEE) 1–8

[54] Jeihouni P, Dehzangi O, Amireskandari A, Rezai A and Nasrabadi N M 2021 *Gan-based super-resolution and segmentation of retinal layers in optical coherence tomography scans 2021 IEEE Int. Conf. on Image Processing (ICIP)* (Piscataway, NJ: IEEE) 46–50

[55] Man N, Guo S, Yiu K F C and Leung C K S 2023 *Multi-layer segmentation of retina oct images* via *advanced U-Net architecture Neurocomputing* **515** 185–200

[56] Cai S, Tian Y, Lui H, Zeng H, Wu Y and Chen G 2020 *Dense-U-Net: a novel multiphoton in vivo cellular image segmentation model based on a convolutional neural network Quant. Imaging Med. Surg.* **10** 1275

[57] Szegedy C, Liu W, Jia Y, Sermanet P, Reed S, Anguelov D, Erhan D, Vanhoucke V and Rabinovich A 2015 *Going deeper with convolutions Proc. of the IEEE Conference on Computer Vision and Pattern Recognition* (Piscatatway, NJ: IEEE) 1–9

[58] Oktay O, Schlemper J, Le Folgoc L, Lee M, Heinrich M, Misawa K, Mori K, McDonagh S, Hammerla N Y, Kainz B *et al* 2018 *Attention U-Net: learning where to look for the pancreas arXiv preprint* arXiv:1804.03999

[59] He K, Zhang X, Ren S and Sun J 2016 *Deep residual learning for image recognition Proc. IEEE Conf. on Computer Vision and Pattern Recognition* (Piscataway, NJ: IEEE) 770–8

[60] Alom M Z, Yakopcic C, Hasan M, Taha T M and Asari V K 2019 *Recurrent residual U-Net for medical image segmentation J. Med. Imaging* **6** 014006–6

[61] Zhou Z, Rahman Siddiquee M M, Tajbakhsh N and Liang J 2018 *U-Net++: a nested U-Net architecture for medical image segmentation Deep Learning in Medical Image Analysis and Multimodal Learning for Clinical Decision Support: 4th Int. Workshop, DLMIA 2018, and 8th Int. Workshop, ML-CDS 2018* (Berlin: Springer) 3–11

[62] Hu J, Shen L and Sun G 2018 *Squeeze-and-excitation networks Proc. IEEE Conf. on Computer Vision and Pattern Recognition* (Piscataway, NJ: IEEE) 7132–41

[63] Kugelman J, Allman J, Read S A, Vincent S J, Tong J, Kalloniatis M, Chen F K, Collins M J and Alonso-Caneiro D 2022 *A comparison of deep learning U-Net architectures for posterior segment oct retinal layer segmentation Sci. Rep.* **12** 14888

[64] Farshad A, Yeganeh Y, Gehlbach P and Navab N 2022 *Y-Net: a spatiospectral dual-encoder network for medical image segmentation Medical Image Computing and Computer Assisted Intervention–MICCAI 2022: 25th Int. Conf.* (Berlin: Springer) 582–92

IOP Publishing

Image Processing with Python
A practical approach
Irshad Ahmad Ansari and Varun Bajaj

Chapter 6

Image denoising using wavelet thresholding technique in Python

Devanand Bhonsle, Shruti Tiwari, Roshni Rahangdale, Manjushree Nayak, Ruhi Uzma Sheikh and Anu G Pillai

Image de-noising is an essential field in image processing, encompassing a wide range of applications. This is a pre-processing task in which unwanted noise signals are removed using different techniques. Noise is unwanted signals that deteriorate the useful information from the image. This information may be edges, ridges, contours or other fine structures. For different applications these details are important. Noise signals may contaminate the image partially or completely. It depends upon the type of noise and its level. Noise may be categorised according to its characteristics. The most frequent types of noise signals encountered in image processing include additive white Gaussian noise, speckle noise, salt and pepper noise, Rician noise, random noise, and more. Noise signals introduced in the images during data acquisition, transmission or due to faulty location. Additive white Gaussian noise is one of the most common noise signals that affects almost all the images to a certain extent. In this chapter we apply a de-noising technique which is based on wavelet thresholding. A wavelet transform is widely recognized as one of the most popular transforms in signal and image processing. It is used in various image processing applications. Thresholding is an essential component in wavelet transforms, and it is commonly classified into two types: hard thresholding and soft thresholding. In this chapter we apply a soft thresholding technique that outperforms the hard thresholding technique.

6.1 Introduction

In our day-to-day life, images are important for various reasons. Almost all mobile users have an inbuilt camera in their mobile phone. Images find applications in various domains, including law enforcement, remote sensing, satellite imaging, surveillance, medical imaging, and many more [1]. In this section we will discuss all the important application areas in image processing, types of noise signals and various medical imaging techniques. Image processing is a significant research area that attracts the attention of researchers due to its importance and wide range of applications, and because it increases the human interpretation, i.e., information can be extracted and processed from the images so that it can be used in various applications such as traffic sensing technologies, law enforcement, remote sensing, weather forecasting, medical science, biometrics, video processing, etc [2]. However, images are corrupted by various types of noise signals. To interpret the images correctly, it is essential to suppress these noise signals [3, 4].

6.2 Literature review

To understand the concept of image de-noising, various research papers have been studied, of which some of the wavelet-based research has been discussed here. Much research has been done in this field. Many researchers have explored and discussed various methods to effectively reduce or suppress different types of noise in images, such as additive white Gaussian noise [5], speckle noise, Rician noise, and impulsive noise. White Gaussian noise contaminates almost all the frequency bands, and speckle noise [1, 6] is the inherent property of those imaging techniques which use coherent signals [7]. For example, ultrasound imaging is one of the most common imaging techniques and it uses ultrasound signals, which is a coherent source. Salt and pepper noise is a common example of impulsive noise that can appear in images as random white and black dots [8, 9]. It may be introduced due to faulty location or error during transmission. Rician noise is a type of noise commonly encountered in magnetic resonance imaging (MRI). It is characterized by its particular statistical distribution. Several research papers have focused on exploring wavelet-based methods [10, 11] for denoising in the context of Rician noise in MRI.

Pan *et al* (1999) proposed two wavelet based de-noising methods [10]. First is improvement of conventional noise filtration techniques in the spatial domain proposed by Chang *et al* (2000), who proposed a data-driven soft thresholding method which is derived in a Bayesian framework hence named 'Bayes-Shrink' [11]. Sveinsson and Benediktsson (2002) used DWT-based oversampled filter banks to reduce the effect of speckle noise from SAR images. These filter banks are called double density DWT [12]. Pizurica *et al* (2003) proposed a noise removal technique from US and MRI images. It is a wavelet domain technique that can adapt itself to various noise signals [13]. Chen *et al* (2004) developed a wavelet thresholding technique that incorporates the neighbouring coefficients, hence this method is called 'Neigh-Shrink' [14]. Dengwen and Xiaoliu (2008) proposed a thresholding technique called 'Block-Shrink' to de-noise images. It is a completely data-driven technique that uses relevancy of neighbor wavelet coefficients using a block thresholding technique [15]. Zhang and Gunturk (2008) implemented a non-linear denoising technique that combines wavelet thresholding and a bilateral filter (BF). It is referred as a multi-resolution bilateral filter. It does the averaging in the spatial domain without smoothing edges [16]. Luisier and Blu (2008) suggested a vector or matrix extension of their de-noising algorithm for the color images, which was developed for grayscale images [17]. It is based on Stein's Unbiased Risk Estimate (SURE) approach [18]. Zhang *et al* (2009) proposed a wavelet transform-based image de-noising algorithm. It is called the M-Bayes threshold, which combines multiple wavelet representations with adaptive thresholding called 'Bayes thresholding' [19]. Hassan and Saparon (2011) proposed an image de-noising method to remove Gaussian noise [20] using DWT. DWT can capture the energy of any signal in a few energy transform values. Hard or soft thresholding can be used with a translation invariant method to overcome the artifacts problems. However, performance of this method can be improved further by using an adaptive shrinkage function [21]. Kongo *et al* (2012) presented a de-noising technique to remove speckle noise from US images using a DTCWT approach. Speckle degrades the quality of low-contrast lesions, which affects medical interpretation and diagnosis processes [22]. In this study, the performance of DTCWT is mixed with bivariate shrinkage and Visu-shrink thresholding techniques. The result demonstrates that DTCWT-based adaptive thresholding techniques are more effective than classical DWT-based methods. Iqbal *et al* (2013) proposed a DTCWT-based non-local means (NLM) method for resolution enhancement of satellite images. The non wavelet-based resolution enhancement scheme loses high frequency contents, which causes blurring, while a DWT-based resolution enhancement scheme generates artifacts because of shift-variant property [23].

From the above literature survey it is clear that wavelet-based denoising techniques remove noise signals efficiently.

6.3 Methodology

After conducting an extensive literature survey, it can be concluded that wavelet-based techniques have demonstrated superior performance in removing additive white Gaussian noise (AWGN) and speckle noise. However, it fails to remove impulsive noise. In this section, wavelet-based techniques have been discussed. This technique has three simple steps only; here is a concise description of the denoising process using wavelet-based techniques:

1. Perform the wavelet transform on the noisy image.
2. Apply a thresholding method to the coefficients obtained from the wavelet transform.
3. Reconstruct the denoised image by performing the inverse wavelet transform using the thresholded coefficients.

By following these steps, the wavelet-based approach effectively removes noise from the image, resulting in a denoised version with enhanced visual quality. However the performance depends upon thresholding techniques and its parameter [6, 24].

6.3.1 Wavelet transform

Wavelet transforms are particularly useful for signals that cannot be adequately represented by the fast Fourier transform (FFT) or other transforms [25]. Unlike symmetrical and periodic sinusoidal functions that extend infinitely in time, wavelets are non-symmetrical and irregular in nature [26]. They are capable of representing signals that contain harmonics, oscillations, and localized features. Wavelets are essentially concentrated waves in both time and space, possessing finite energy. This unique property makes them suitable for capturing localized information in signals and images, which is not achievable using other transform methods. WT is used to analyze the transient signals. The major advantage of a wavelet is that it can represent the signal without any loss of information contained in it. It provides not only good time resolution but frequency resolution also. It is broadly divided into two categories; continuous wavelet transform (CWT) and discrete wavelet transform (DWT) [27].

6.3.2 Discrete wavelet transform

Discrete wavelet transform (DWT) is a discrete and decomposition-based form of wavelet transform [27]. It is the discrete and decomposition form of CWT, which provides two types of coefficients with each decomposition viz. approximation and detailed coefficients. DWT is given as:

$$\psi_{m,n}(t) = \frac{1}{\sqrt{|2^m|}}\psi\left(\frac{t-2^m n}{2^m}\right) \qquad (6.1)$$

where m and n are the scaling and translation index at each decomposition level.

In the process of DWT, the input signal is passed through a low pass filter (LPF) followed by a downsampler to obtain the approximation coefficients. Similarly, passing the input through a high pass filter (HPF) followed by a downsampler yields the detail coefficients. This decomposition can be performed for multiple levels, where the coefficients obtained at each level serve as input for the subsequent level. It is important to note that as the number of levels increases, the resolution of the signal tends to deteriorate. DWT is an invertible transform, meaning that the original signal can be reconstructed using the obtained coefficients. This is achieved through the inverse discrete wavelet transform (IDWT). By applying the IDWT process, the original signal can be reconstructed using the coefficients obtained at each level of decomposition. Wavelet decomposition and wavelet reconstruction form the basis of the DWT, allowing for efficient analysis and processing of signals at different scales or resolutions.

6.3.3 Wavelet decomposition

As previously mentioned, in the process of DWT, an input signal undergoes two steps to obtain approximation and detailed coefficients. First, the signal is passed through an LPF and then downsampled, resulting in the approximation coefficients. Second, the signal is passed through an HPF and downsampled, yielding the detailed coefficients. This process is performed separately for each dimension (rows and columns) of the data. For a 2D grayscale image [17], one decomposition step involves splitting the original image into four sub-images, each with a size of $N/2 \cdot N/2$. These sub-images contain information from different frequency components.

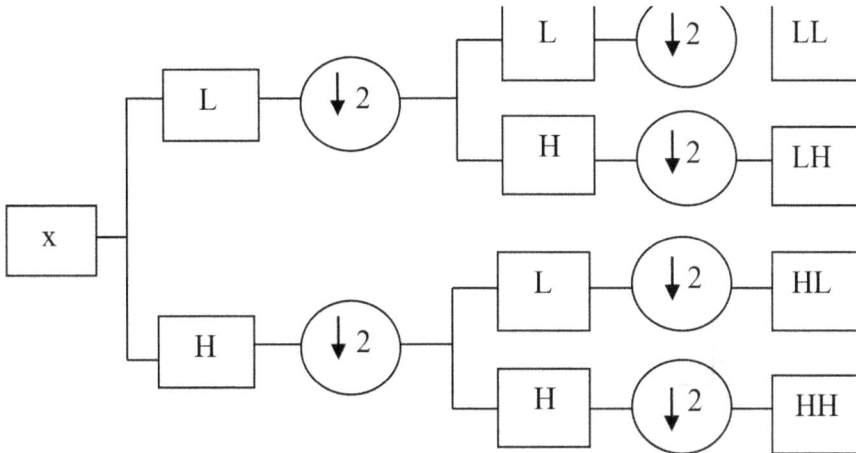

Figure 6.1. One decomposition step of a 2D image.

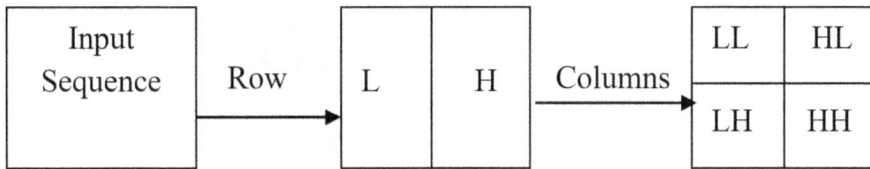

Figure 6.2. One-level DWT decomposition.

Figure 6.1 illustrates one decomposition step of the 2D grayscale image, showing the split into four sub-images. Figure 6.2 depicts the four sub-bands resulting from the one-level decomposition. This decomposition process enables the analysis of image data at different frequency components, facilitating applications such as image compression, denoising, and feature extraction.

In the process of DWT decomposition, the resulting sub-bands have specific characteristics:

- LL sub-band: It is obtained by applying an LPF in both the row and column directions. This sub-band contains a coarse or rough description of the image and is referred to as the approximation sub-band.
- HH sub-band: It is obtained by applying an HPF in both the row and column directions. This sub-band contains high-frequency components, representing fine details in the image.
- LH and HL sub-bands: These sub-bands are obtained by applying LPF in one direction (horizontal) and HPF in the other direction (vertical) or vice versa. The LH sub-band captures vertical details corresponding to horizontal edges, while the HL sub-band captures horizontal details corresponding to vertical edges. These two sub-bands are referred to as the detail sub-bands.

Figure 6.3 demonstrates a three-level DWT decomposition, illustrating the resulting sub-bands at each level. The decomposition process can be performed iteratively on

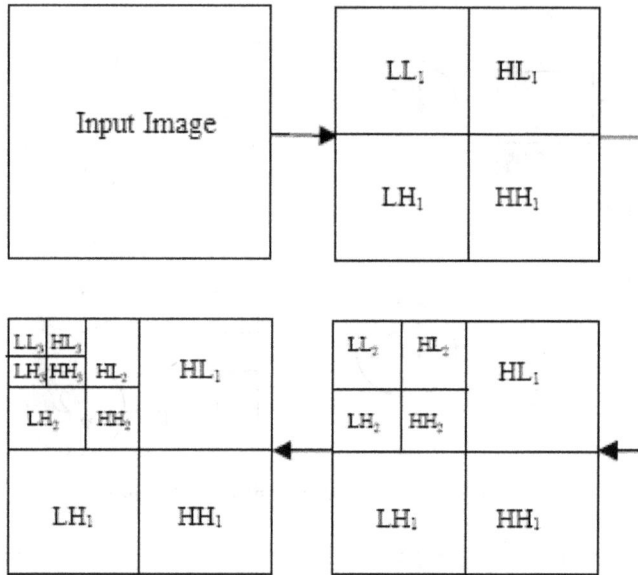

Figure 6.3. Three-level DWT decomposition.

an image, allowing for further analysis of different frequency components. The number of decompositions can be adjusted based on the specific requirements of the application or the desired level of signal representation.

6.3.4 Wavelet reconstruction

The IDWT is the reverse process of the decomposition, used to reconstruct the original signal using wavelet coefficients. In IDWT, all the sub-bands are up-sampled and filtered using corresponding inverse filters along the columns. The results that belong together are then added, up-sampled, and filtered again with the corresponding inverse filters. The asterisk (*) denotes the inverse filter operation. The result of the last stage is added together to obtain the reconstructed original image, x. By applying the IDWT process, the original signal can be reconstructed using the wavelet coefficients obtained during the decomposition. This allows for the retrieval of the original image from its wavelet representation, facilitating signal recovery and restoration.

When a 2D image undergoes decomposition using DWT and is subsequently reconstructed, there is no loss of information, and the exact original image is obtained. Figure 6.4 illustrates the reconstruction process, showcasing how the original image can be reconstructed from the wavelet coefficients.

DWT is widely recognized as an effective tool in various application areas, with image processing being one of its most common and prominent applications. In image processing tasks, DWT often outperforms many other non-wavelet techniques in terms of denoising, compression [11], feature extraction, and other related

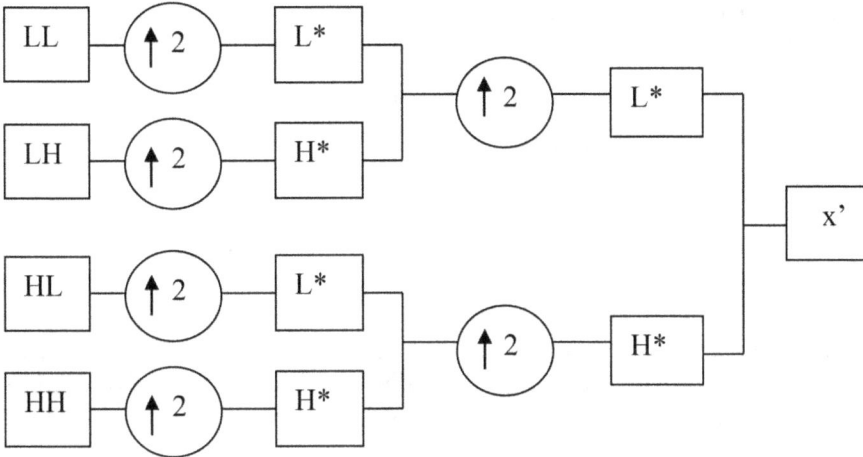

Figure 6.4. One composition step of four sub-images.

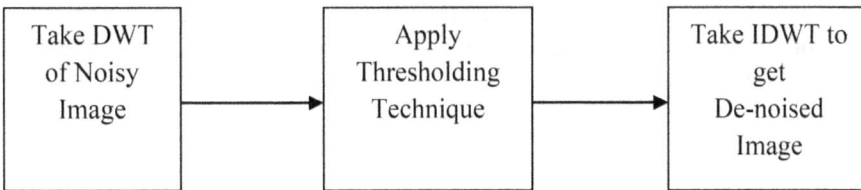

Figure 6.5. DWT based image de-noising technique.

tasks. Its ability to capture both frequency and spatial information efficiently contributes to its superior performance in image processing tasks.

6.3.5 Wavelet based de-noising technique

The DWT is a powerful tool widely used in various digital image processing applications. It enables the simultaneous analysis of signals in both the time and frequency domains. With its versatility, the DWT finds extensive applications in image processing tasks. Image denoising is one such application area that involves manipulating image data to produce a high-quality, noise-free image. The process of image denoising using DWT typically involves three steps:

- Calculate the DWT of the noisy image signal.
- Modify the wavelet coefficients of the noisy signal based on specific rules or techniques.
- Perform the IDWT to obtain the denoised image.

Figure 6.5 depicts the block diagram of a DWT-based image denoising technique, illustrating the flow of the denoising process using wavelet transforms. By utilizing DWT for image denoising, it becomes possible to effectively reduce noise and enhance the quality of images.

Consider an image x that has been corrupted by noise n, resulting in a noisy image w. Mathematically, this can be represented as:

$$w = x + n \tag{6.2}$$

Taking DWT of equation (3.15) we get,

$$W = X + N \tag{6.3}$$

In the first step, DWT of the noisy image is calculated. In the second step, any shrinkage technique is applied on it, which modifies the noisy coefficients using some rule. This process is called 'thresholding' and in the last step IDWT is calculated which provides a de-noised image [2]. The efficiency of the de-noising technique strongly depends on the types of thresholding applied on the noisy coefficients. Thresholding may be broadly divided into two categories as mentioned below:

6.3.6 Hard thresholding

In this particular scheme, a threshold value (λ) needs to be determined. The input is left unchanged if it is greater than λ; otherwise, it is set to zero. This thresholding rule is applied to the wavelet coefficients of the detailed sub-bands, while the low-resolution coefficients remain unaltered. The hard thresholding operator can be expressed as:

$$W = T_{hard}(X,\ W) = \begin{cases} X, |X| \geqslant \lambda \\ 0, |X| < \lambda \end{cases} \tag{6.4}$$

Using this operator, the wavelet coefficients in the detailed sub-bands that exceed the threshold λ are preserved, while those below the threshold are set to zero, effectively suppressing the noise in the detailed sub-bands. The low-resolution coefficients are not affected by this thresholding operation.

Hard thresholding is an effective tool to retain edge and boundary information present in the image. Figure 6.6 illustrates the hard thresholding scheme.

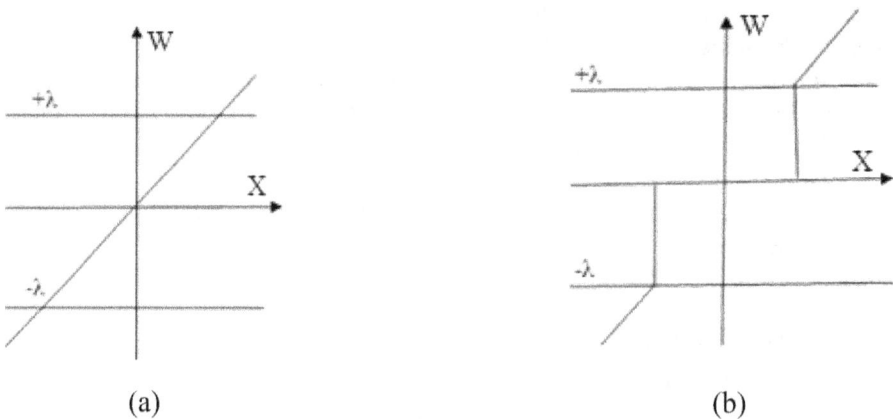

(a) (b)

Figure 6.6. Hard thresholding scheme (a) original signal (b) after hard thresholding.

6.3.7 Soft thresholding

In this scheme, if the coefficient has a magnitude greater than the selected threshold value then it is shrunk towards zero otherwise set to zero. The soft-thresholding operator may be given as (figure 6.7):

$$W = T_{soft}(X, W) = \begin{cases} \text{sgn}\{X\}(|X| - \lambda), |X| \geqslant \lambda \\ 0 \qquad '|X| < \lambda \end{cases} \qquad (6.5)$$

Soft thresholding outperforms hard thresholding in terms of image denoising. Figure 6.7 illustrates the soft thresholding scheme. Bivariate thresholding is the example of adaptive soft thresholding, which outperforms many hard thresholding and soft thresholding techniques.

6.3.8 Wavelet families

There are many wavelets, each characterized by its amount of symmetry, vanishing moments, support width and regularity.

6.3.8.1 Haar wavelets

The Haar wavelet is one of the earliest and simplest wavelets used in wavelet analysis. It has a step-like shape, with a value of 1 on the interval [0, 1/2] and a value of −1 on the interval [1/2, 1]. The Haar wavelet possesses a single vanishing moment, which means it can effectively represent and capture constant or linear components in signals. Its support width is one, indicating that it is localized to a single interval. To compute the Haar transform, running averages and differences are calculated using scaling signals and wavelets. This process involves performing a sequence of operations to obtain coefficients that represent the signal at different scales and positions. Figure 6.8 provides a visual representation of the Haar wavelet, illustrating its step-like shape and characteristic features [3].

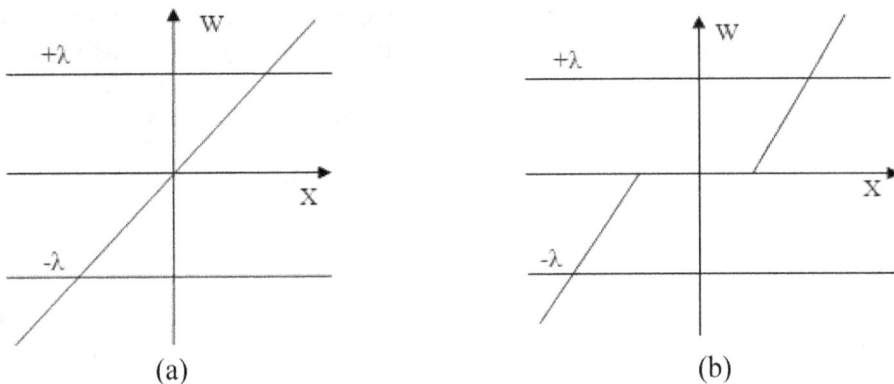

(a)   (b)

Figure 6.7. Soft thresholding scheme (a) original signal (b) after soft thresholding.

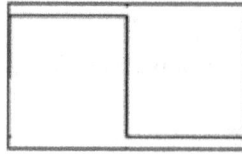

Figure 6.8. The Haar wavelet.

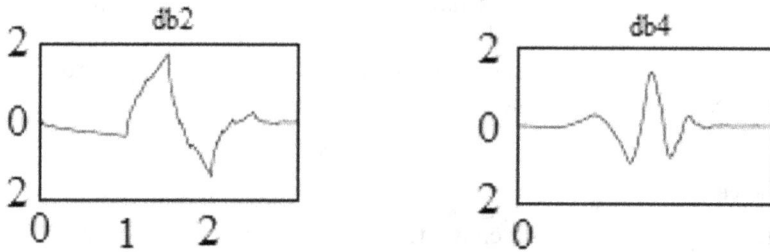

Figure 6.9. Daubechies wavelets.

6.3.8.2 Daubechies wavelets

The Daubechies wavelets, named after Ingrid Daubechies, are a well-known family of wavelets in the field of wavelet research. They are defined in a similar manner to the Haar transform, involving the computation of running averages and differences. However, the Daubechies transform employs different scaling signals and wavelets for these computations, resulting in enhanced capabilities compared to the Haar transform. The support of Daubechies wavelets is larger, allowing for the calculation of averages and differences using more signal values. This expansion of support width significantly improves the performance and effectiveness of these transforms. Daubechies wavelets offer a wide range of applications, including compression and noise removal in audio signals and images. The Daubechies family of wavelets is characterized by compactly supported wavelets, with the highest number of vanishing moments relative to their support width. These wavelets are identified by their name (e.g., db or Daub) followed by the order, which denotes the support width. For example, db1 is similar to the Haar wavelet, while dbN has a support width of $2N-1$ and N vanishing moments. It's worth noting that some authors may use $2N$ instead of N to denote the order of Daubechies wavelets. Daubechies wavelets are not symmetric and exhibit distinct characteristics. Figure 6.9 provides a visual representation of a series of Daubechies wavelets, highlighting their diverse shapes and properties.

6.3.8.3 Coiflets

The family of Coiflets wavelets, also designed by Ingrid Daubechies based on suggestions from Coifman, is characterized by compact support and aims to maintain a close correspondence between trend values and the original signal values. Coiflets wavelets are constructed as an extension of the Daubechies wavelets.

Coiflets have a support width of $6N-1$, providing a wider range of analysis compared to Daubechies wavelets. They possess the highest number of vanishing moments for both the scaling signal ($2N-1$) and the wavelet ($2N$). Some authors may use $2N$ instead of N to denote the order of these wavelets. In contrast to Daubechies wavelets, Coiflets wavelets exhibit greater symmetry. They are more balanced in terms of their positive and negative components, resulting in improved performance and symmetry properties. Figure 6.10 showcases a series of Coiflets wavelets, illustrating their diverse shapes and characteristics.

6.3.8.4 Symlets

Symlets, as the name suggests, are wavelets that are designed to be nearly symmetrical. They were proposed by Ingrid Daubechies as a modification of the Daubechies wavelets. Symlets are characterized by compact support, with a support width of $2N-1$, and they have N vanishing moments. In comparison to the original Daubechies wavelets, Symlets exhibit increased symmetry, which can be advantageous in various signal processing applications. Their nearly symmetrical nature allows for improved representation and analysis of signals with both positive and negative components. Figure 6.11 showcases a series of Symlets wavelets, highlighting their different shapes and characteristics.

6.3.8.5 Mexican hat

This wavelet has no scaling function. It is derived from a function proportional to the second derivative function of the Gaussian probability density function (figure 6.12).

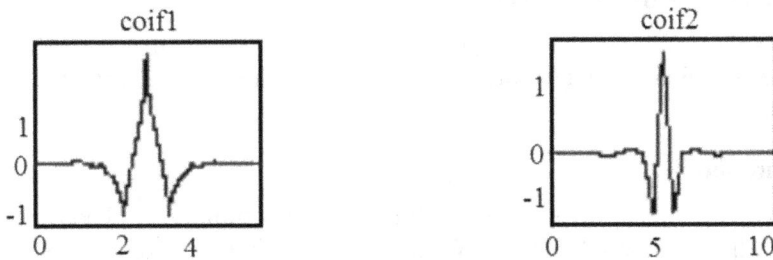

Figure 6.10. Coiflets.

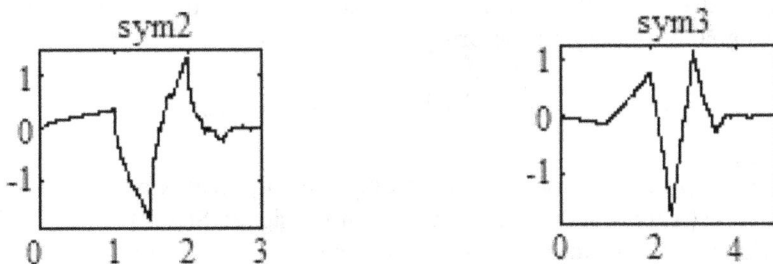

Figure 6.11. Symlets.

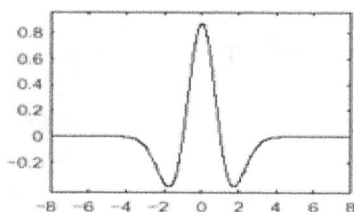

Figure 6.12. Mexican hat.

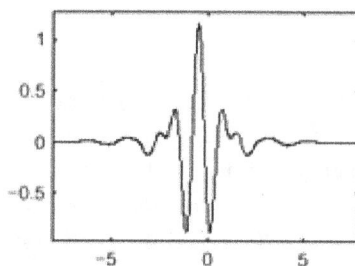

Figure 6.13. Meyer wavelet.

6.3.8.6 Meyer

The Meyer wavelet and scaling function are defined in the frequency domain (figure 6.13).

6.4 Thresholding selection

The list of six thresholding techniques are as follows: Sure Shrink, Bayes Shrink [28], Bivariate Shrinkage, Neigh Shrink, Block Shrink and Non Parametric statistical approach.

6.4.1 Sure shrink

Donoho and Johnstone proposed a threshold selection method known as Sure Shrink, which utilizes Stein's Unbiased Risk Estimator (SURE). Sure Shrink combines the concepts of the universal threshold and the SURE threshold. For each resolution level j in the wavelet transform, Sure Shrink determines a level-dependent threshold value t_j. The main objective of Sure Shrink is to minimize the mean squared error (MSE), which is a measure of the discrepancy between the denoised signal estimate and the original noise-free signal.

$$\text{MSE} = \frac{1}{n^2} \sum_{X,Y-1}^{n} (Z(X, Y) - S(X, Y))^2 \tag{6.6}$$

Sure Shrink suppresses noise by thresholding the empirical wavelet coefficients. In this method, $Z(X, Y)$ represents the estimate of the signal, $S(X, Y)$ corresponds to the original signal without noise, and n denotes the size of the signal. The threshold t^* for Sure Shrink is defined as follows:

$$t^* = \min\left(t,\ \sigma\sqrt{2\log n}\right) \tag{6.7}$$

An estimate of the noise level σ is defined based on the median absolute deviation, given by the expression. Here, t represents the value that minimizes Stein's Unbiased Risk Estimator, and σ denotes the noise variance.

$$\hat{\sigma} = \frac{\text{median}(\{|g_{j-1,\,k}|: k = 0,\ 1,\ \ldots,\ 2^{j-1} - 1\})}{0.6745} \tag{6.8}$$

It is worth noting that n represents the size of the image. Moreover, Sure Shrink is a smoothness adaptive method, implying that if the unknown function within the image contains abrupt changes or boundaries, these features will also be captured in the reconstructed image [24, 26].

6.4.2 Bayes shrink

Chang, Yu, and Vetterli introduced Bayes Shrink, aiming to minimize the Bayesian risk, which led to its name, Bayes Shrink. The Bayes threshold, denoted as t_B, is defined as follows:

$$t\mathrm{B} = \frac{\sigma^2}{\sigma_s} \tag{6.9}$$

The noise variance is represented as σ^2, while the signal variance without noise is denoted as σ_s. Considering the definition of additive noise as $W(x, y) = s(x, y) + n(x, y)$, where $W(x, y)$ represents the noisy signal, $s(x, y)$ represents the signal without noise, and $n(x, y)$ represents the independent noise component.

$$\sigma_w^2 = \sigma_s^2 + \sigma^2 \tag{6.10}$$

σ_w^2 can be computed as shown below:

$$\sigma_w^2 = \frac{1}{n^2}\sum_{X,\,Y-1}^{n} w^2(X,\ Y) \tag{6.11}$$

The variance of the signal σ_s^2 is computed as:

$$\sigma_s = \sqrt{\max\left(\sigma_w^2 - \sigma^2,\ 0\right)} \tag{6.12}$$

With σ^2 and σ_s^2, the Bayes threshold is computed from equation (6.12). Using this threshold the wavelet coefficients are threshold at each band.

6.5 Python code for image de-noising

```
sigma_est=estimate_sigma()
# Denoising is to be done using wavelet transform
X=denoise_wavelet()
# Two soft thresholding technique is to be used namely
```

Bayes Shrink and Vishushrink
```
# Number of level used for decomposition is 3
Y= denoise_wavelet(x, wavelet= 'Wname', mode= 'SorH',
wavelet_level =n, method= 'BayesShrink', rescle_sigma=
'TorF'
Y= denoise_wavelet(x, wavelet= 'Wname', mode= 'SorH',
wavelet_level =n, method= 'visuShrink', rescle_sigma=
'TorF'
```

Python Code for Grayscale Image Denoising

```
# Denoising of gray scale image has been done using
skimage
# Images are corrupted by Additive White Gaussian Noise
(awgn)

import matplotlib.pyplot as plt
from skimage.restoration import (denoise_wavelet, esti-
mate_sigma)
from skimage.util import awgn noise
from skimage.metrics import peak_signal_noise_ratio
import skimage.io

# Read image
img=skimage.io.imread ('stones.jpg)

# Convert image as float
img=skimage_as_float(img)

#Add awgn with standard deviation 0.01
sigma=0.01 # noise std
imgn=awgn_noise(img, var=sinma**2)

sigma_est=estimate_sigma(imgn,       average_sigmas=True)
noise estimation

# Denoise using Bayes Shrink

img_bayes=      denoise_wavelet      (imgn,      method=
'BayesdShrink', mode= 'soft', wavelet_levels=3, wave-
let='bior6.8, rescale_sigma=True)

# Denoise using Visu Shrink
```

```
img_visushrink=    denoise_wavelet    (imgn,    method=
'visuShrink', mode= 'soft', sigma=sigma_est/3 wavele-
t_levels=5, wavelet='bior6.8, rescale_sigma=True)

#find Power Signal to Noise Ratio(psnr)

psnr_noisy=peak_signal_noise_ratio (img, imgn)
psnr_bayes= peak_signal_noise_ratio (img, imgn_bayes)
psnr_   visushrink   =   peak_signal_noise_ratio   (img,
imgn_visushrink)

#print PSNR

print ('PSNR[Original vs noisy iamage]:, psnr-noisy)
print ('PSNR[Original vs denoisy iamage]:, psnr-noisy)
print ('PSNR[Original vs denoised(Bayes)]:, psnr-noisy)
```

6.6 Results and discussion

To validate the performance of the aforementioned denoising technique it is desire to apply noise in the noise free images and then denoising technique is to be applied on noisy images. The results of the denoising method are presented through images and peak signal-to-noise ratio (PSNR) values. Specifically, the denoising of grayscale images is demonstrated below. Figure 6.14(a) displays the original image, while figure 6.14(b) represents the image corrupted with noise. Figures 14(c) and (d)

(a) Original Image

(b) Noisy Image

(c) Denoised Image using Bayes Shrink

(d) Denoised Image using visu Shrink

Figure 6.14. Input and output gray images.

Table 6.1. PSNR values for gray images for different values of AWGN (Daubechies 3).

Gray Image	PSNR	AWGN Noise level					
		0.01	0.02	0.05	0.1	0.2	0.5
Stones	Noisy PSNR	23.1302	20.4359	18.8167	13.6146	13.1454	12.7007
	PSNR_Visu	34.6758	34.2079	33.9043	29.7762	28.8126	28.2873
	PSNR_Bayes	35.0584	34.4402	34.0015	30.2765	29.6754	28.6541
Flowers	Noisy PSNR	22.8302	20.5259	18.6432	13.7175	13.1765	12.7245
	PSNR_Visu	35.476	34.248	34.0333	30.9831	29.0134	26.5642
	PSNR_Bayes	35.5691	35.1101	34.9871	31.8712	30.5432	27.9871
lena	Noisy PSNR	23.6444	20.5432	18.9873	13.8687	13.1123	12.6218
	PSNR_Visu	35.2156	34.1403	34.4532	30.7685	28.6329	26.5104
	PSNR_Bayes	35.524	35.0296	34.8406	31.128	29.994	27.0986

Table 6.2. PSNR values for color images for different values of AWGN (Symlet 4).

Color Image	PSNR	AWGN Noise level					
		0.01	0.02	0.05	0.1	0.2	0.5
Stones	Noisy PSNR	25.1232	22.7812	20.7878	15.5677	13.6667	13.0124
	PSNR_Visu	36.5444	35.9143	32.6571	30.3176	27.7634	27.5541
	PSNR_Bayes	37.2342	36.3241	33.7658	31.4545	28.0154	27.3247
Flowers	Noisy PSNR	24.4545	21.6572	19.9983	14.0231	13.6667	13.5645
	PSNR_Visu	35.3333	34.2343	33.7682	31.0123	29.6571	25.9812
	PSNR_Bayes	36.3452	35.6543	34.2234	32.7854	30.8763	26.1023
lena	Noisy PSNR	23.7845	20.7786	19.6574	14.0234	13.0231	12.8794
	PSNR_Visu	36.5463	35.9512	34.7868	31.5554	29.8763	27.1287
	PSNR_Bayes	37.0231	36.4532	35.5445	32.4382	30.4351	28.3333

showcase the denoised images obtained using Bayes Shrink and Visu Shrink methods, respectively.

Table 6.1 shows PSNR (in dB) values of noisy and denoised images for gray images using db3 wavelets.

Table 6.2 shows PSNR(in dB) values of noisy and denoised images for gray images using symlet 4 wavelets.

References

[1] Andria G, Attivissimo F, Lanzolla A and Savino M 2013 *A suitable threshold for speckle reduction in ultrasound images IEEE Trans. Instrum. Meas.* **62** 2270–9

Image Processing with Python

[2] Bhonsle D, Chandra V K and Sinha G R 2015 Gaussian and speckle noise removal from ultrasound images using bivariate shrinkage by dual tree complex wavelet transform *i-manager's J. Image Process.* **2** 1–5

[3] Bhonsle D 2021 Denoising of digital images using wavelet-based thresholding techniques: a comparison cognitive behavior and human computer interaction based on machine learning algorithm *Cognitive Behavior and Human Computer Interaction Based on Machine Learning Algorithm* (New York: Wiley) 85–115

[4] Bhonsle D 2021 Quality improvement of Richardson lucy based de-blurred images using krill herd optimization *IEEE Int. Conf. on Advances in Electrical Computing Communication and Sustainable Technologies* (Piscataway, NJ: IEEE) 1–5

[5] Gonzalez R C and Woods R E 2009 *Digital Image Processing* 3rd edn (Englewood Cliffs, NJ: Prentice-Hall)

[6] Dewangan N and Bhonsle D 2013 Comparison of wavelet thresholding for image denoising using different shrinkage *Int. J. Emer. Trends Technol. Comput. Sci.* **2** 57–61

[7] Bhonsle D, Sinha G R and Chandra V K 2020 Medical image de-noising using combined bayes shrink and total variation techniques *Artificial Intelligence and Machine Learning in 2D/3D Medical Image Processing* (Boca Raton, FL: CRC Press) pp 31–52

[8] Bhonsle D, Sinha G R and Shrivastava D P 2020 Suppression of Gaussian noise using spatio-spectral total variation technique *7th IEEE Int. Conf. on Information Technology Trends* (Piscataway, NJ: IEEE) 89–93

[9] Bhonsle D and Dewangan S 2012 Comparative study of dual-tree complex wavelet transform and double density complex wavelet transform for image denoising using wavelet-domain *Int. J. Sci. Res Pub.* **2** 1–5

[10] Chandrakar N and Bhonsle D 2013 A new hybrid image denoising method using bilateral filter and DWT *Int. J. IT Eng. Appl. Sci. Res.* **2**

[11] Bhonsle D, Chandra V K and Sinha G R 2018 De-noising of CT Images using combined bivariate shrinkage and enhanced total variation technique *i-Manager's J. Electron. Eng.* **8** 12

[12] Tiwari A and Bhonsle D 2016 Salt and pepper noise estimation and removal techniques and its performance evaluation *i-Manager's J. Pattern Recognit.* **3** 28

[13] Tomasi C and Manduchi R 1998 Bilateral filtering for gray and color images *IEEE Int. Conf. Computer Vis.* (Piscataway, NJ: IEEE) 839–46

[14] Bhonsle D, Chandra V K and Sinha G R 2018 An optimized framework using adaptive wavelet thresholding and total variation technique for de-noising medical images *J. Adv. Res. Dyn. Control Sys* **10** 953–65

[15] Bhonsle D, Bagga J, Mishra S, Sahu C, Sahu V and Mishra A 2022 Reduction of Gaussian noise from computed tomography images using optimized bilateral filter by enhanced grasshopper algorithm *2022 Int. Conf. on Advances in Electrical, Computing, Communication and Sustainable Technologies (ICAECT)* (Piscataway, NJ: IEEE) 1–9

[16] Bhonsle D, Rizvi T, Mishra S, Sinha G R, Kumar A and Jain V K 2022 Reduction of ultrasound images using combined bilateral filter & median modified wiener filter *2022 Int. Conf. on Advances in Electrical, Computing, Communication and Sustainable Technologies (ICAECT)* (Piscatawat, NJ: IEEE) 1–6

[17] Pan Q, Zhang L, Dai G and Zhang H 1999 Two denoising methods by wavelet transform *IEEE Trans. Signal Process.* **47** 3401–6

[18] Chang S G, Yu B and Vetterli M 2000 Adaptive wavelet thresholding for image denoising and compression *IEEE Trans. Image Process.* **9** 1532–46

[19] Chen G Y, Bui T D and Krzyzak A 2004 Image denoising using neighbouring wavelet coefficients *IEEE Int. Conf. on Acoustics, Speech, and Signal Processing* (Piscataway, NJ: IEEE) 917–20

[20] Pizurica A, Philips W, Lemahieu I and Acheroy M 2003 A versatile wavelet domain noise filtration technique for medical imaging *IEEE Trans. Med. Imaging* **22** 323–31

[21] Sveinsson J R and Benediktsson J A 2002 Double density wavelet transformation for speckle reduction of SAR images *IEEE Int. Geoscience and Remote Sensing Symp. Toronto* (Piscataway, NJ: IEEE) 113–5

[22] Dengwen Z and Xiaoliu S 2008 Image denoising using block thresholding *IEEE Congress on Image and Signal Processing (Sanya, Hainan, China)* pp 335–8

[23] Zhang M and Gunturk B K 2008 Multiresolution bilateral filtering for image denoising *IEEE Trans. Image Process.* **17** 2324–33

[24] Luisier F and Blu T 2008 SURE-LET Multichannel image denoising: interscale orthonormal wavelet thresholding *IEEE Trans. Image Process.* **17** 482–92

[25] Zhang W, Yu F and Guo H M 2009 Improved adaptive wavelet threshold for image denoising *IEEE Chinese Control and Decision Conf.* (Piscataway, NJ: IEEE) 5958–63

[26] Kongo R M, El-Kadmiri M L O, Hassanain N and Cherkaoui Md 2012 Dual-tree complex wavelet in medical ultrasounds images restoration *IEEE Int. Conf. on Multimedia Computing and Systems* (Piscataway, NJ: IEEE) 297–303

[27] Hassan H and Saparon A 2011 Still image denoising based on discrete wavelet transform *IEEE Int. Conf. on System Engineering and Technology* (Piscataway, NJ: IEEE) 188–91

[28] Iqbal Z, Ghafoor A and Siddiqui A M 2013 Satellite image resolution enhancement using dual-tree complex wavelet transform and nonlocal means *IEEE Geosci. Remote Sens. Lett.* **10** 451–5

IOP Publishing

Image Processing with Python
A practical approach
Irshad Ahmad Ansari and Varun Bajaj

Chapter 7

Prostate cancer segmentation of peripheral zone and central gland regions in mpMRI: comparative analysis with deep neural network U-Net and its advanced models

Anil B Gavade, Rajendra B Nerli, Pushkar Bansidhar Patil, Richa Ravi Siddannavar, Venkata Siva Prasad Bhagavatula and Priyanka A Gavade

Prostate cancer, a highly prevalent and significant health concern affecting men worldwide, necessitates timely detection and effective treatment. Genetic and environmental factors contribute to its development, emphasizing the importance of regular screenings and genetic testing for early diagnosis. Multiparametric magnetic resonance imaging (mpMRI) has shown great potential as a diagnostic tool for prostate cancer, enabling the identification and characterization of cancerous lesions within the peripheral zone (PZ) and central gland (CG) of the prostate. This paper proposes a novel approach utilizing deep learning-based segmentation and classification algorithms for prostate cancer analysis. The segmentation task accurately delineates the PZ and CG regions in mpMRI volumes, while the classification task differentiates between benign and malignant prostate tissues. Comparative evaluations of various deep learning architectures, including U-Net, 3D U-Net, V-Net, and nnU-Net, revealed that nnU-Net exhibited superior performance in accurately segmenting prostate cancer within the PZ and CG regions. Furthermore, the comparative analysis of deep learning-based classification frameworks highlighted ResNet-50 as the leading model with the highest accuracy. These findings demonstrate the capabilities of deep learning techniques, specifically nnU-Net for segmentation and ResNet-50 for classification, to improve the precision and efficiency of prostate cancer diagnosis and treatment planning using mpMRI, ultimately enhancing patient outcomes.

doi:10.1088/978-0-7503-5924-5ch7

7.1 Introduction

The prostate gland plays a vital role in the functioning of the male reproductive system, characterized by its small, walnut-shaped structure. The prostate gland is situated beneath the bladder and positioned anteriorly to the rectum. Describing the size of a healthy adult prostate gland involves its dimensions and weight. On average, a normal adult prostate gland typically measures approximately 3–4 centimeters in length, 3–4 centimeters in width, and 2–3 centimeters in height. The weight of a healthy prostate gland usually ranges from 20–30 grams. Its primary function revolves around the production and secretion of fluid, an essential component of semen. This fluid serves the purpose of nourishing, safeguarding, and facilitating the transportation of semen during ejaculation. Additionally, the prostate gland encompasses the urethra, which is responsible for urine passage, and its muscular contractions aid in urine flow. Given its location, the prostate gland has a crucial role in the functions related to male reproduction and the urinary system.

Various factors such as age, hormonal imbalances, obesity, genetic factors, inflammation, lifestyle choices, race, ethnicity, and exposure to certain toxic environments have been suggested as potential causes of prostate gland diseases, although their direct relationship remains unproven. When the prostate gland enlarges, it can exert pressure on the urethra, resulting in the obstruction of urine flow. This can lead to urinary symptoms such as urgency, weak urine flow, difficulty initiating urination, and in some cases, urinary incontinence. It is worth noting that certain neurological conditions can also contribute to urinary incontinence, which occurs when there is an inability to control the flow of urine effectively. Importantly, not all men exposed to these risk factors develop prostate diseases, while some men without exposure to these risk factors may still experience prostate gland diseases. Therefore, regular screening, early detection, and appropriate medical care play a crucial role in managing male health and addressing any underlying conditions.

Various prostate gland diseases consist of benign prostatic hyperplasia (BPH), prostatitis, prostate cancer, prostatic intraepithelial neoplasia (PIN), prostate abscess, and prostate stones. Prostate cancer is characterized by the malignant growth of cells in the prostate gland and is one of the most prevalent cancers among men. Although prostate cancer may not exhibit symptoms in its early stages, it can lead to urinary symptoms, blood in the urine or semen, erectile dysfunction, and bone pain as it progresses. According to the International Agency for Research on Cancer (IARC), prostate cancer is the cancer that receives the second highest number of diagnoses and ranks as the fifth leading cause of cancer-related fatalities among men worldwide. Global estimates for 2020 indicate around 1.4 million new cases of prostate cancer and approximately 3 75 000 deaths. In India, prostate cancer constitutes about 7%–10% of all cancer cases in men, and its incidence is on the rise due to changing lifestyles, improved diagnostic techniques, and increased awareness.

The prostate gland is situated deep within the pelvis, just below the bladder, and surrounds the urethra. Due to its small size, similarity to neighboring tissues, natural variations in size among individuals, and motion artifacts caused by bladder and bowel movements during breathing, imaging the prostate gland presents challenges.

X-ray imaging is insufficient for detailed visualization of soft tissues, while ultrasound imaging has limitations in capturing the posterior part of the gland. MRI is widely regarded as an effective imaging technique, although it has constraints related to patient positioning and image interpretation. PET imaging can reveal the spread of malignant tissue, and biopsy serves as another diagnostic tool for detecting cancers. Considering its ability to provide detailed soft tissue information, MRI is commonly used as the initial clinical investigation. Specialized imaging techniques such as transrectal ultrasound and multiparametric MRI (mpMRI) are employed for prostate imaging. In the context of this chapter, mpMRI images are utilized for the study of prostate gland images. The proposed architecture for image segmentation and classification are shown in figure 7.1.

Numerous deep-learning approaches have been extensively studied in order to tackle the difficulties associated with analyzing MRI images and distinguishing between different types of prostate tissue. These techniques primarily concentrate on accurately segmenting the gland and precisely classifying malignant tissue. mpMRI encompasses a range of sequences, such as T2-weighted images, which aid in visualizing the anatomy of the prostate, including the central zone, peripheral zone, and transition zone, allowing for the detection of abnormalities or lesions. Another sequence is diffusion weighted imaging (DWI), which provides insights into the movement of water molecules within the prostate tissue, aiding in cellular understanding. Derived from DWI, the apparent diffusion coefficient (ADC) mapping quantifies water molecule movement, and lower ADC values associated with higher cellularity may indicate the presence of malignant growth. Additionally, dynamic contrast-enhanced (DCE) imaging is employed to visualize blood flow to the tissue, helping detect increased blood flow and categorize lesions. The rich information present in multiparametric MRI images necessitates the use of algorithms that can accurately segment the gland despite its challenging location and size, ensuring high accuracy in classification.

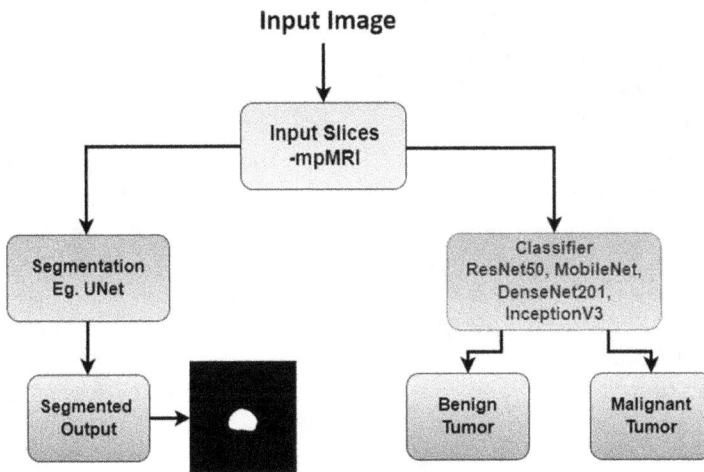

Figure 7.1. mpMRI image cancer segmentation and classification.

7.2 Literature reviews

Hassan *et al* [1] proposed a novel automated classification algorithm using deep learning approaches for prostate cancer detection from ultrasound and MRI images. The maximum accuracy achieved on US images was 97%, while for MRI images it was 80%, and performance was further improved through fusion with shallow machine learning algorithms.

Mirasol *et al* [2] implemented a repeatable framework using U-Net with batch normalization to separate prostate cancer lesions from annotated apparent diffusion coefficient maps, achieving a Dice score of 0.47 (0.44–0.49).

Fernandez-Quilez [3] explored DL algorithms tailored for limited data scenarios in prostate cancer MRI applications, including generative adversarial networks (GAN) for segmentation and lesion significance classification, and auto-encoders (AEs) to address data imbalance. The proposed AE-based framework achieved results of 73.90 DSC and 0.79 AUC.

Zhao *et al* [4] developed deep learning models on multicenter bpMRI to accurately detect prostate cancer, surpassing the performance of PIRADS assessment by expert radiologists in terms of AUC and specificity, except in one external validation cohort.

Daneshvar *et al* [5] utilized a DL detection and grading system on mpMRI of active surveillance (AS) patients to assess lesion dynamics and correlate findings with progression. Progressors showed increased high-risk lesions and disease burden, and patients with persistent high-risk lesions had a higher risk of progression.

Zaridis *et al* [6] proposed a DL-based cropping pipeline using U-Net to improve prostate PZ segmentation. Cropping significantly improved segmentation performance across all three networks, measured by the Dice score, Hausdorff distance, sensitivity and average surface distance, with improvements ranging from 13% to 34%.

Chun *et al* [7] evaluated the performance of CNN algorithms (ResNet50, InceptionV3, VGG16). Prostate cancer detection was performed on 620 image samples obtained from TCIA data source. The accuracy rankings for the tested models were as follows: VGG16 achieved the highest accuracy with a rate of 95.56%, ResNet50 at 86.67%, and InceptionV3 at 85.56%.

Mehralivand *et al* [8] aimed to create a cascaded deep learning system that utilizes biparametric prostate MRI and PI-RADS for the detection and classification of prostate abnormalities, with the aim of assisting radiologists. The algorithm achieved 56.1% sensitivity, 62.7% PPV, and a false discovery rate of 37.3%. The DSC median score for segmentation of lesion was 0.359., and the classification accuracy of PI-RADS was 30.8%.

Ye *et al* [9] introduced a deep learning network called PSP-Net+VGG16 for diagnosing prostate tumors, which achieved a segmentation Dice index of 91.3%. The classification accuracy and recognition rates using VGG16 were 87.95% and 87.33% correspondingly.

Hosseinzadeh *et al* [10] designed a DL framework to speculate PI-RADS$\geqslant$4 lesions and Gleason>6 lesions from bpMRI. The DL achieved a sensitivity of 87%

for PI-RADS⩾4 lesion detection with an average of 1 false positive (FP) per patient and an AUC of 0.88. The sensitivity for Gleason>6 lesion detection was 85% @ 1 FP compared to a consensus panel of expert radiologists with 91% @ 0.3 FP.

Hu *et al* [11] involved 396 prostate cancer patients and created four deep learning models: PM1, PM2, AM1, and AM2. The models using apparent diffusion coefficient and T2-weighted images outperformed the pretraining models (PM1 and PM2) in both internal validation and test datasets, achieving higher AUCs and κ values for prostate cancer classification and grading.

Koonamparampath *et al* [12] summarized the usage of CNN in various automatic processing tasks for prostate cancer detection and diagnosis. Deep learning-based research outperformed traditional patient prognosis techniques in terms of accuracy.

Roest *et al* [13] used a multi-center dataset of 1513 patients who underwent bpMRI. Training the machine learning classifier on both prior and current study data showed higher diagnostic accuracy (AUC 0.81) compared to using the current study only (AUC 0.73). Adding clinical variables further improved performance (AUC 0.86).

Bleker *et al* [14] used a multi-center study of 524 prostate cancer lesions on bpMRI, the DLM auto-fixed VOI method achieved a significantly higher AUC (0.76) compared to manual segmentation (0.62).

Pellicervalero *et al* [15] used two datasets, the study achieved excellent results for lesion-level AUC, sensitivity, and specificity in mpMRI testing. The model outperformed radiologists in both lesion-level and patient-level evaluation.

Salvi *et al* [16] developed an automated hybrid method for MR image-based prostate gland segmentation, achieving excellent results with a mean dice score of 0.851 and a Hausdorff distance of 7.55 mm.

Künzel *et al* [17] evaluated autonomous radiotherapy planning for prostate cancer cases by five experts, with acceptance rates ranging from 66% to 90% for different steps.

Fan *et al* [18] included 252 patients who received MP-MRI tests. Among six classifiers, RF performed best for different diagnostic groups, and SVM performed best for a specific group. DCE features ranked first in each group's models in terms of number and importance.

Quihui-Rubio *et al* [19] observed that among six deep learning models trained on MRI images, the R2U-Net model achieved the highest scores for segmenting all zones: Dice (0.869), Jaccard (0.782), and mean squared error (0.000 13).

Jing *et al* [20] observed that a radiomics signature combining whole prostate and lesion features, along with PI-RADS, outperformed subjective evaluation alone in predicting clinically significant prostate cancer, as confirmed by ROC analysis, NRI, and DCA.

Banerjee *et al* [21] created an ensemble comprising of three convolutional layers and two fully connected layers, which yielded a 72% accuracy and an AUC of 0.76 when evaluated on a dataset consisting of 40 patients. Additionally, their Q-learning agent attained an accuracy of 73% and an AUC of 0.73.

Adams *et al* [22] used 158 expert-annotated prostate MRIs. The segmentation models achieved the following scores: central gland—DSC/HD/ASD of 0.88/18.3/ 2.2, peripheral zone—0.75/22.8/1.9, and PCa—0.45/36.7/17.4.

Li *et al* [23] developed a multi-modal CNN called RMANet for discriminating prostate cancer clinical severity grade. Their model achieved an AUC of 0.84, outperforming other classical and recent methods.

Tyagi *et al* [24] utilized deep learning and computer vision in their research to improve pathology computer-aided diagnosis systems. They demonstrated the effectiveness of their approach on a fully annotated dataset.

Twilt *et al* [25] provided an overview of AI algorithms used for identifying and categorizing lesions in prostate cancer. They found heterogeneity in cohort sizes and limited evidence to strongly support the clinical effectiveness of AI applications for prostate cancer (PCa).

Cipollari *et al* [26] reviewed 170 mpMRI scans. Inter-reader agreement was moderate, and the PCa detection rate was similar between an experienced and inexperienced reader.

Hu *et al* [27] conducted a retrospective investigation on 217 patients. The effectiveness of deep convolutional neural networks (DCNN) was enhanced by utilizing transfer learning from disease scans for prostate cancer detection.

Collado [28] used a ResNext101 3D as an encoder and a Unet3D as a decoder for their approach. The U-Net 3D model provided the best results.

Yin [29] investigated machine learning classifiers for each prostate zone. CNNs worked best for the AS zone, while ensemble algorithms worked best for PZ and TZ zones.

Van Sloun *et al* [30] employed deep learning technique for automatic segmentation of the prostate zone and transition zone using datasets from three different machine, and achieved a Jaccard index of 0.93 for automatic segmentation.

These studies highlight the advancements in deep learning approaches for prostate cancer segmentation and classification, demonstrating improved accuracy and performance compared to traditional methods.

7.3 Materials

7.3.1 Dataset details

Algorithm implemented on **Dell Precision Tower 5810 work station** with specification Xeon CPU, 512 GB SSD, 32 GB RAM, 2 TB HDD and 8 GB Quadro P4000 Nvidia GPU. Datasets employed to check the performance of algorithm are as follows.

Promise12 Dataset:
- The PROMISE12 [31] dataset is a freely accessible dataset created specifically for evaluating the effectiveness of prostate magnetic resonance imaging (MRI) segmentation algorithms. The objective is to evaluate the effectiveness of algorithm in accurately delineating the prostate gland and its peripheral zone from multi-parametric MRI scans.
- Comprising T2-weighted MRI scans and corresponding ground truth segmentation masks, the dataset encompasses information obtained from 50

patients diagnosed with prostate cancer. It offers a comprehensive collection of data for researchers to analyze and develop prostate MRI segmentation algorithms.
- Renowned for its utilization in numerous research studies, the PROMISE12 dataset serves as a widely recognized benchmark for assessing the precision and reliability of prostate MRI segmentation algorithms. Researchers rely on this dataset to evaluate and compare the accuracy and robustness of their algorithmic approaches in this domain.

ProstateX Dataset:
- The ProstateX [32] dataset encompasses a compilation of multi-parametric MRI scans and clinical data related to the prostate gland, specifically targeting prostate cancer detection and diagnosis.
- Within this dataset, one can find diffusion-weighted MRI, magnetic resonance spectroscopy data, T2-weighted MRI and dynamic contrast-enhanced MRI. Additionally, it contains relevant information such as prostate volume, prostate-specific antigen (PSA) levels, and indications of the presence or absence of prostate cancer.
- The creation of the ProstateX dataset was part of the ProstateX challenge, which sought to develop computer-aided diagnosis systems based on MRI data for prostate cancer. Consequently, it serves as a significant resource for researchers dedicated to the detection and classification of prostate cancer. The ProstateX dataset is divided into two distinct subsets, each serving a specific purpose:
- Prostate3T dataset: This subset focuses on multi-parametric magnetic resonance imaging (mp-MRI) scans obtained using a 3 Tesla (3T) MRI scanner. It includes diffusion-weighted MRI, T2-weighted MRI, and dynamic contrast-enhanced MRI scans of the prostate. The primary objective of the Prostate 3T dataset is to facilitate the development and evaluation of computer-aided diagnosis (CAD) algorithms for detection of prostate cancer. While the specific size and contents of this subset may vary, it typically consists of a substantial number of patient cases.
- ProstateDx dataset: This subset is dedicated to the diagnosis of prostate cancer using biopsy information. It comprises data from patients who underwent a biopsy after the mp-MRI scans. The ProstateDx dataset provides crucial details regarding the presence or absence of cancer in different biopsy cores. Its primary role lies in evaluating and validating the performance of CAD algorithms for prostate cancer diagnosis.

7.3.2 Reference repository details
Segmentation models:
1. U-Net: The GitHub repository [33] contains an implementation of U-Net specifically designed for the PROMISE12 challenge. The challenge focuses on prostate segmentation in MRI images. The code available in the

repository allows for the training and evaluation of the U-Net model on the PROMISE12 dataset.

2. 3D U-Net: The GitHub repository [34] implements the 3D U-Net architecture for prostate segmentation of 3D MRI volumes. The repository includes the code for preprocessing the data, training the model, and evaluating the segmentation performance. It also provides pre-trained models and example usage scripts.

3. V-Net: The V-Net architecture used for segmentation of medical images, can be found in a GitHub repository called VNet [35]. This repository provides an implementation of the V-Net model along with code for training and testing on different datasets. It also provides pre-trained models, evaluation scripts, and instructions for usage.

4. nnU-Net: The GitHub repository [36] is dedicated to nnU-Net, an advanced deep learning framework developed for segmenting medical images. This repository provides a comprehensive collection of code, instructions, and example notebooks to facilitate the utilization of nnU-Net for various medical image segmentation tasks. It encompasses essential components such as pre-processing techniques, model training procedures, and inference pipelines.

Classification models:

For medical image classification endeavors, the GitHub repository offers an assortment of purpose-built classification models [37]. These models are specifically designed to cater to the unique requirements of medical image classification tasks. The repository comprises of well-documented code for training and evaluating convolutional neural network (CNN)-based models on medical image datasets. Additionally, it provides pre-trained models, utilities for data preprocessing, and example scripts to facilitate usage and customization.

7.4 Methodology

7.4.1 Segmentation models

- **U-Net**

  U-Net [38] is a deep learning framework utilized for image segmentation purposes, particularly in the biomedical domain. The original research paper titled 'U-Net: Convolutional Networks for Biomedical Image Segmentation,' published by Ronneberger *et al* in 2015, introduced this architecture (figure 7.2).

  The primary concept behind U-Net involves employing a fully convolutional neural network to directly generate a segmentation mask for input images. Unlike conventional methods that depend on pre-processing steps, feature extraction, and classification, U-Net is capable of performing all these tasks seamlessly by learning from the data.

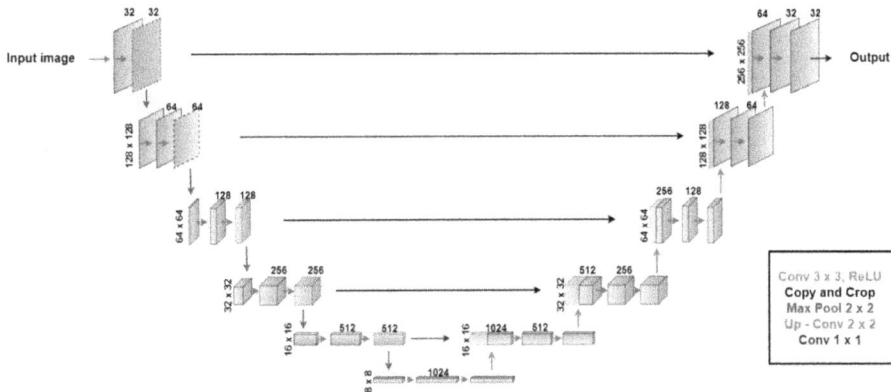

Figure 7.2. U-Net architecture.

The U-Net architecture comprises of the given components:
1. **Contracting Path (Encoder):** The U-Net architecture follows an encoder-decoder design. The encoder path captures the contextual information of the image through multiple layers of 3×3 convolutional filters, combined with 2×2 max pooling operations to extract features and reduce resolution. To ensure non-linearity, a rectified linear unit (ReLU) activation function is applied after every convolutional layer in an end-to-end manner.
2. **Expanding Path (Decoder):** Conversely, the decoder path utilizes up-sampling layers and 3×3 convolutional filters to reconstruct the segmentation mask from the features obtained in the encoder path. Each up-sampling layer is paired with a 2×2 transpose convolutional layer that doubles the feature resolution.
3. **Skip Connections:** Skip connections are created between the corresponding layers of the encoder and decoder paths in order to restore the spatial information that is lost during the encoding process. These connections concatenate the feature maps from both paths, enabling the decoder to utilize the spatial information from the encoder.
4. **Output Layer:** The final layer of the U-Net architecture is a 1×1 convolutional layer, which generates a probability map for each pixel in the input image. This map indicates the likelihood of the pixel belonging to a particular class, facilitating the segmentation task.

Overall, U-Net efficiently handles large images while accurately performing segmentation, making it suitable for a variety of applications such as cell segmentation, road segmentation, and biomedical image segmentation. The inclusion of skip connections allows U-Net to handle objects with diverse sizes and shapes by preserving spatial information throughout the network.

- **3D-Unet**

    3D U-Net [39], a variation of the U-Net architecture, was proposed by Çiçek *et al* in 2016 as a solution for volumetric image segmentation tasks. It

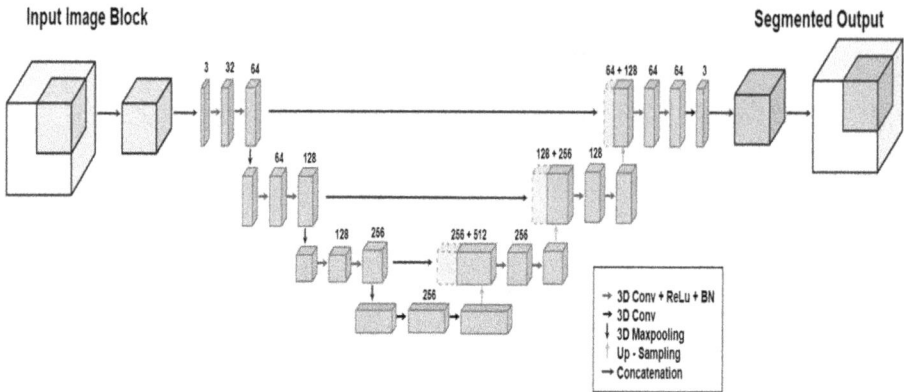

Figure 7.3. 3D U-Net architecture.

specifically targets 3D medical images, which are represented as stacks of 2D slices. By directly producing a segmentation mask for 3D input volumes, 3D U-Net proves advantageous in medical image analysis. The model architecture is shown in figure 7.3.

The structure of 3D U-Net closely resembles the original U-Net but with adjustments to handle 3D volumes. Notably, it employs 3D convolutional filters instead of 2D filters to enable effective processing of the volumetric data.

The 3D U-Net architecture encompasses the following components:

1. **Contracting Path (Encoder):** This stage involves multiple layers of $3\times3\times3$ convolutional filters, followed by $2\times2\times2$ max pooling operations, for feature extraction and resolution reduction. The presence of a rectified linear unit (ReLU) activation function is observed in every convolutional layer.

2. **Expanding Path (Decoder):** The features obtained from the encoder path undergo up-sampling layers and $3\times3\times3$ convolutional filters to reconstruct the segmentation mask. Each up-sampling layer is followed by a $2\times2\times2$ transpose convolutional layer, doubling the feature resolution.

3. **Skip Connections:** Skip connections are used in 3D U-Net, similar to U-Net, to connect corresponding layers of the encoder and decoder paths. These connections concatenate feature maps along the depth, height, and width dimensions, facilitating the recovery of spatial information lost during the contracting path.

4. **Output Layer:** The last layer comprises of a $1\times1\times1$ 3D convolutional layer that generates a probability map for each voxel in the input volume, indicating the likelihood of voxel classification.

Overall, 3D U-Net presents a robust architecture for 3D volumetric image segmentation, particularly in medical image analysis. It has proven effective in various segmentation tasks. By utilizing 3D convolutional filters, 3D U-Net captures spatial information in all three dimensions, a crucial aspect for accurate segmentation in volumetric images.

- **V-Net**

  V-Net [40] is a deep learning framework specifically developed for 3D image segmentation purposes. The architecture, proposed in the research article titled 'V-Net: Fully Convolutional Neural Networks for Volumetric Medical Image Segmentation' by Milletari *et al* in 2016, employs a fully convolutional neural network to directly generate a segmentation mask for 3D input volumes.

  Unlike traditional methods that rely on a combination of preprocessing, feature extraction, and classification, V-Net has the capability to learn and execute all these tasks seamlessly in an end-to-end manner. V-Net follows an encoder-decoder design, similar to U-Net. However, instead of employing skip connections to merge encoder and decoder features, V-Net utilizes a 3D convolutional residual network (ResNet) to facilitate information propagation throughout the network. The model architecture is shown in figure 7.4.

  In detail, the V-Net architecture encompasses the following components:
  1. **Encoding Path:** The input volume undergoes multiple layers of 3D convolutional filters and max pooling operations to extract features and decrease resolution.
  2. **Residual Blocks:** A series of residual blocks, built upon 3D convolutional layers, are employed to learn deep representations of the input volume. Each residual block comprises two convolutional layers with a skip connection that bypasses the convolutional layers.

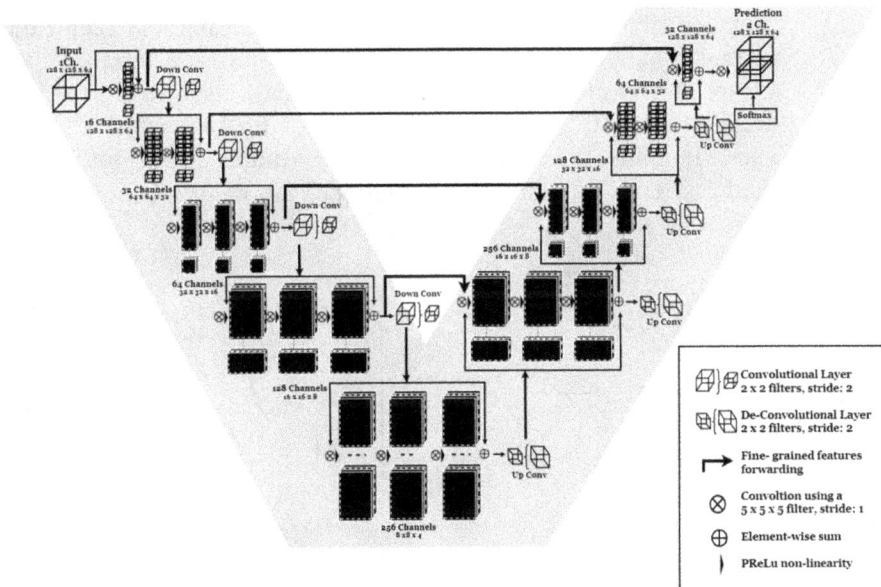

Figure 7.4. VNet architecture.

3. **Decoding Path:** The features obtained from the encoding path are fed through a series of up sampling layers and 3D convolutional filters to reconstruct the segmentation mask. Each up sampling layer is followed by a 3D deconvolutional layer that enhances the feature resolution.

4. **Output Layer:** The last layer of this network consists of a 1×1×1 3D convolutional layer, that produces a probability map for every voxel in the input volume. This map signifies the likelihood of the voxel belonging to a specific class.

The V-Net architecture excels in efficiently and accurately handling large 3D medical images, making it highly suitable for diverse medical image segmentation tasks.

- **nnU-Net**

The nnU-Net (No New-Net) [41] is a deep learning framework explicitly designed for medical image segmentation. Isensee *et al* introduced this approach in their paper 'nnU-Net: Self-adapting Framework for U-Net-based Medical Image Segmentation' published in 2021 (figure 7.5).

nnU-Net aims to provide a versatile and efficient solution for segmentation of medical images by incorporating adaptations to the well-known U-Net architecture. The core concept of nnU-Net revolves around adaptability to different segmentation tasks and datasets. It comprises of several key components, including preprocessing, the U-Net architecture, training strategy, and postprocessing.

The nnU-Net architecture is composed using the following components:

1. **Preprocessing**: In the preprocessing stage, input medical images undergo normalization and resizing to ensure consistent size and resolution. This step contributes to enhanced performance during the segmentation process.

2. **U-Net Architecture:** nnU-Net architecture builds upon the U-Net architecture, introducing modifications to further optimize performance. It splits the contracting and expanding paths into separate modules, each containing convolutional layers, batch normalization,

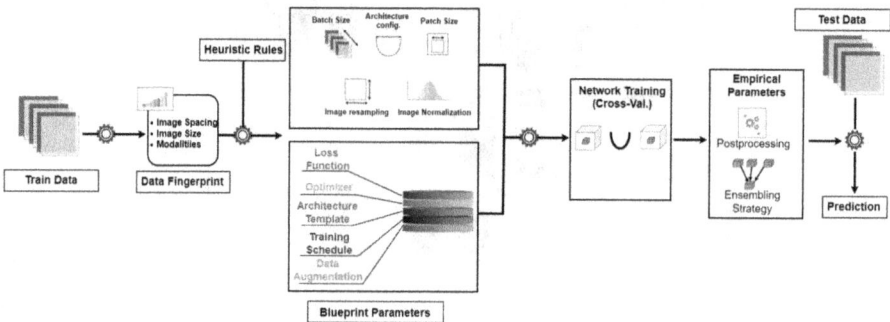

Figure 7.5. nnU-Net architecture.

and activation function ReLU. The number of blocks and filters can be adjusted based on the complexity of the segmentation task.

3. **Training**: The training process follows a multi-stage approach, starting with the utilization of a small subset of the data to initialize the weights of the model. Subsequent stages involve training on progressively larger subsets, allowing fine-tuning and prevention of overfitting. This strategy enhances the model's generalization capabilities.

4. **Postprocessing**: To refine the output segmentation maps, nnU-Net applies postprocessing techniques. This step involves removing small isolated regions and ensuring a smooth and continuous final segmentation map.

In summary, nnU-Net presents a robust and adaptable solution for segmentation of medical images. Its flexibility in handling different datasets and segmentation challenges makes it well-suited for a diverse array of applications in the field of medical imaging. The utilization of a multi-stage training strategy and postprocessing further enhances the model's performance, establishing nnU-Net as a state-of-the-art architecture for medical image segmentation.

7.4.2 Classification models

- **ResNet-50**

  1. ResNet50 [42] revolutionized deep neural networks by introducing residual connections to combat the degradation problem. These skip or shortcut connections allow the network to learn residual mappings instead of exclusively concentrating on learning the intended underlying mapping. The ResNet-50 model architecture is shown in figure 7.6.

  2. The ResNet50 architecture comprises multiple residual blocks, which act as building blocks. Each residual block consists of convolutional layers and a shortcut connection that adds the input to the output. This innovative design effectively addresses the vanishing gradient problem and facilitates training of deep networks.

  3. ResNet50 stands out with its 50-layer depth, surpassing previous architectures like VGG16 or GoogLeNet. This increased depth enables more accurate feature extraction and improved representation learning, leading to exceptional execution on difficult visual recognition tasks.

  4. Evidenced by its remarkable achievement in the ImageNet Large Scale Visual Recognition Challenge (ILSVRC) 2015, ResNet50 surpassed

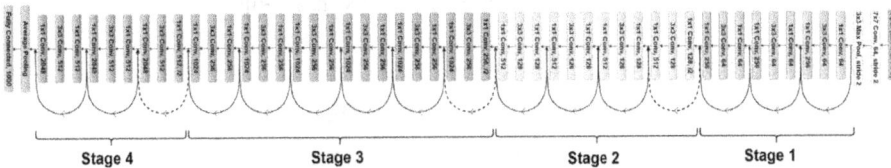

Figure 7.6. ResNet-50 architecture.

human-level accuracy. Consequently, its pre-trained weights have become widely adopted for transfer learning in diverse computer vision applications.

- **MobileNet**
  1. MobileNet [43] is specifically designed to be deployed on devices that have limited computational resources. Its main objective is to achieve a balance between model size, computational efficiency, and accuracy. The MobileNet model architecture is shown in figure 7.7.

  2. One of the key techniques employed by MobileNet is the use of depthwise separable convolutions. These convolutions break down the standard convolution into two parts: depthwise convolution, that operates on individual channels, and pointwise convolution, which combines information across channels. This factorization leads to a significant reduction in parameters and computational cost without sacrificing expressive power.
  3. By leveraging depthwise separable convolutions, MobileNet enables efficient inference on devices with restricted resources, making it well-suited for real-time applications.
  4. It is worth noting that MobileNet comes in various versions, such as MobileNetV1, MobileNetV2, and MobileNetV3, each introducing enhancements in terms of accuracy and efficiency.

- **DenseNet201**
  1. DenseNet201 [44] is an expanded version of the original DenseNet architecture, designed to enhance the sharing and flow of features across layers (figure 7.8).

  2. It employs direct connections between every layer in a feed-forward manner, resulting in dense connectivity that addresses the issue of vanishing gradients and promotes stronger feature propagation.

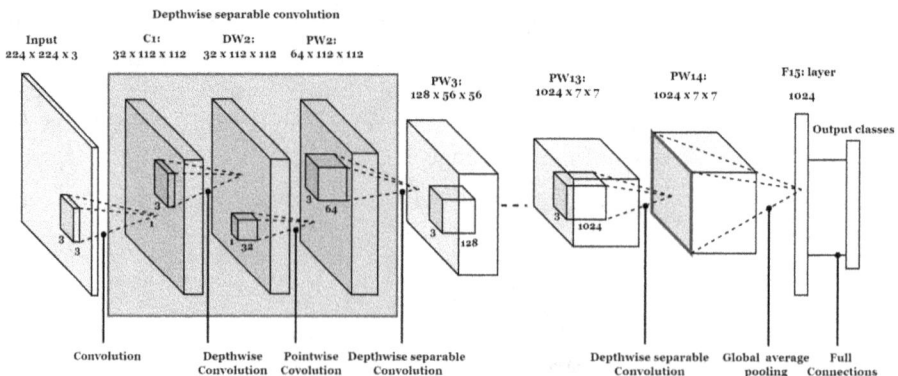

Figure 7.7. MobileNet architecture.

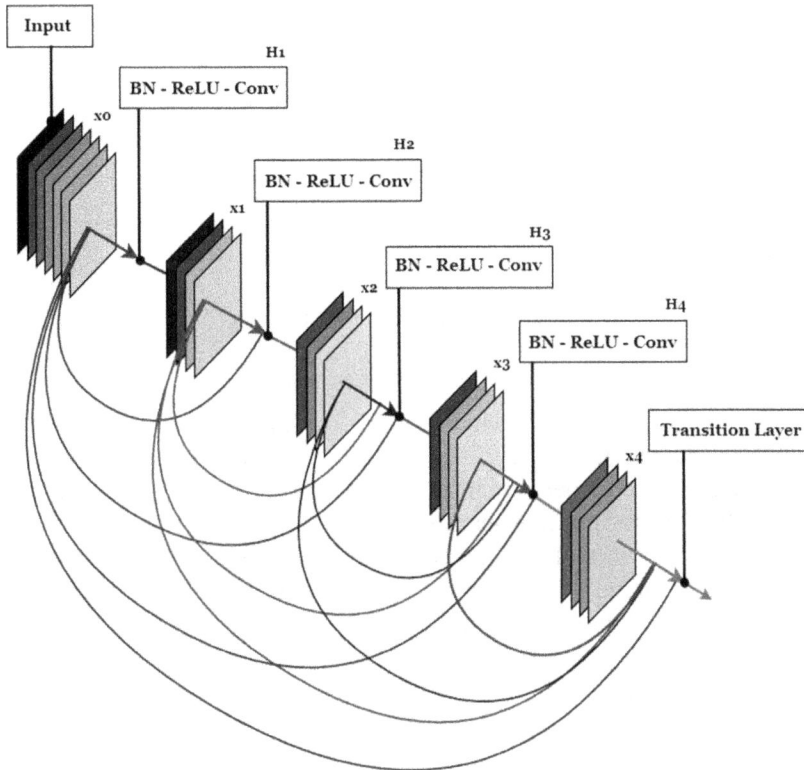

Figure 7.8. DenseNet201 architecture.

3. To facilitate faster convergence and boost performance, DenseNet201 incorporates rectified linear unit (ReLU) activations and batch normalization. Notably, DenseNet201 demonstrates impressive performance in image classification tasks, even with fewer parameters compared to alternative architectures.

4. Its notable strengths include its ability to capture intricate features and acquire discriminative representations.

- **InceptionNetV3**

    1. InceptionNetV3 [45], also known as Inception V3, is an enhanced version of the original Inception architecture designed to enhance computational efficiency while maintaining high accuracy levels. The InceptionNetV3 model architecture is shown in figure 7.9.

    2. It introduces several innovative architectural elements, such as factorized convolutions, which substitute large filters with multiple smaller filters. This substitution reduces computational complexity without compromising performance.

    3. To address challenges like the vanishing gradient problem and to provide regularization benefits, InceptionNetV3 incorporates auxiliary

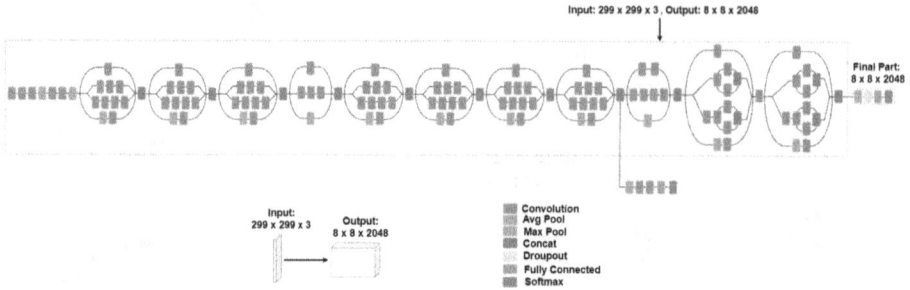

Figure 7.9. InceptionNetV3 architecture.

classifiers at intermediate layers during training. These additional classifiers contribute to stabilizing the training process and improving overall model performance.

4. One of the notable features of InceptionNetV3 is the utilization of inception modules. The modules are comprised of parallel convolutional operations utilizing varying kernel sizes. By incorporating multiple parallel operations, the network can effectively capture information at various scales, enabling the learning of both local and global features simultaneously.

7.4.3 Algorithm

Algorithm 1. Cancer cell detection and classification.   **Input**: mpMRI Image

      Output: Indicate predicted cancer classification

Step 1: Dataset ($\mathbf{R}$) contains $\mathbf{X}$ images and $\mathbf{Y}$ labels, where $\mathbf{X} \in \mathbf{R}$ and $\mathbf{Y} \in \mathbf{R}$

Step 2: Training set ($\mathbf{T_r}$) is 80% of $\mathbf{R}$. i.e., $\mathbf{T_r = 0.80\ R}$

Step 3: Test set ($\mathbf{T_s}$) is 20% of $\mathbf{R}$. i.e., $\mathbf{T_s = 0.20\ R}$

Step 4: Select batch size ($\mathbf{B}$) such that $\mathbf{B} = \{1, 2, 3, 4....\}$ and number of epochs ($\mathbf{E}$)

Step 5: **For** i $= 1$ to $\mathbf{E}$:

1. | Randomly select $\mathbf{xy} = \{x_1y_1, x_2y_2, x_3y_3, ..., x_ny_n\}$ such that $|\mathbf{xy}| = \mathbf{B}$ and $\mathbf{xy} \subseteq \mathbf{T_r}$

2. | Give $\mathbf{xy}$ as input to the model

3. | Save the model if there is increment in accuracy and validation accuracy

4. | **End** for

Step 6: Calculate and display Dice coefficient (for segmentation), confusion matrix and other metrices (for classification) using $\mathbf{T_s}$

7.4.4 Evaluation metrics

The confusion matrix is an important resource utilized in the fields of machine learning and statistics to assess the effectiveness of a classification model. It offers a concise representation of the model's predictions by comparing them against the true labels. Typically represented as a square matrix, the confusion matrix's dimensions are determined by the number of classes in the classification problem. For instance, in a binary classification scenario with classes labeled as positive (P) and negative (N), the confusion matrix would appear as follows (figure 7.10).

Here is a breakdown of the key components within the confusion matrix:

True Positive (TP): These instances occur when the model accurately identifies the positive class (P) while the actual class is also positive.

False Negative (FN): In this scenario, the model mistakenly classifies the negative class (N) despite the actual class being positive.

False Positive (FP): In this case, the model incorrectly predicts the positive class (P) even though the actual class is negative.

True Negative (TN): These cases arise when the model correctly predicts the negative class (N) while the actual class is also negative.

By analyzing the distribution of correct and incorrect predictions across different classes, the confusion matrix enables the calculation and visualization of various evaluation metrics. These metrics include recall (or sensitivity), precision, F1 score and accuracy.

Accuracy: The accuracy metric assesses the model's overall precision, indicating the extent to which it produces correct results and is determined by

$$\text{Accuracy} = \frac{TP + TN}{TP + FP + FN + TN}$$

Precision: This metric expresses the measurement of accurately identifying positive instances among all instances predicted as positive and is determined by

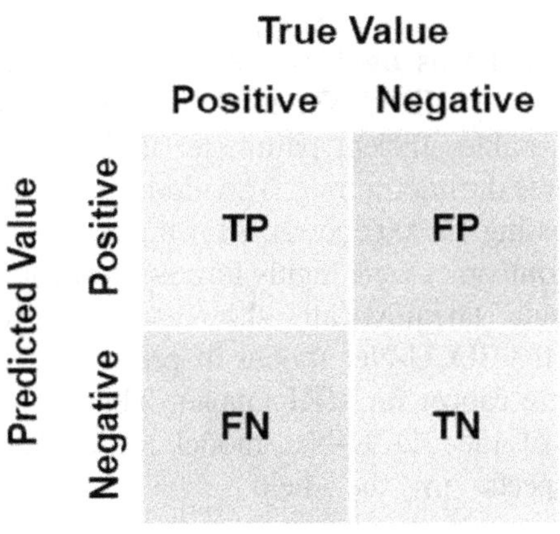

Figure 7.10. Confusion matrix.

$$\text{Precision} = \frac{\text{TP}}{\text{TP} + \text{FP}}$$

Recall: The recall metric measures the accuracy of correctly predicting positive instances out of all the actual positive instances, expressed as a proportion and is determined by

$$\text{Recall} = \frac{\text{TP}}{\text{TP} + \text{FN}}$$

F1 Score: It represents a metric that effectively combines precision and recall by utilizing their harmonic mean. It is computed as twice the multiplication of precision and recall, divided by their sum.

$$\text{F1} - \text{Score} = \frac{2 \times \text{Recall} \times \text{Precision}}{\text{Recall} + \text{Precision}}$$

By utilizing a confusion matrix, analysts and researchers can gain valuable insights into a model's performance and identify areas that require improvement. It is an effective means of evaluating the effectiveness of a classification model.

7.5 Results

7.5.1 Segmentation results

- **U-Net**

  Promise12 dataset was utilized to perform segmentation on mpMRI images of prostate cancer [44]. Using the U-Net model, a deep learning methodology was utilized to accurately outline the regions of interest within the images. The chosen value for the batch size was two and epochs set to 150. After 150 epochs, the following parameters were achieved: a loss of 0.0026, an accuracy of 0.9991, a validation loss of 0.0329, and a validation accuracy of 0.9927. These hyperparameter choices were crucial in achieving the obtained Dice coefficient of 0.913. The algorithm has been tested on the PROMISE12 dataset [44].

- **3D U-Net**

  ProstateX dataset was used as the basis for segmenting prostate cancer mpMRI images using 3D U-Net model. To optimize the model's learning capabilities and enable efficient parameter updates, the chosen value for the batch size was three during training. The model underwent an extensive training process encompassing 15 000 epochs, allowing for iterative improvements in its predictions. The outcomes were highly impressive, achieving a dice coefficient of 0.926. These results unequivocally showcased the exceptional accuracy and performance of the 3D U-Net model in precisely delineating the regions of interest in prostate cancer mpMRI images. This accomplishment underscores the potentiality of the 3D U-Net model for clinical applications, offering promising prospects in the field. The algorithm has been tested on the PROMISE12 dataset NCI-ISBI 2013 challenge dataset [40].

- **V-Net**

  The Promise12 dataset was employed to conduct segmentation on mpMRI images of prostate cancer using the V-Net model. During training, a batch size of four was utilized, allowing the model to process and update its parameters using a larger batch of data simultaneously. The training phase encompassed 100 epochs, ensuring multiple iterations of the entire dataset to enhance the model's predictions. Notably, the outcomes achieved demonstrated exceptional performance, with a Dice coefficient of 0.926. This coefficient signifies a strong agreement amidst the predicted segmentations and the actual ground truth, affirming the V-Net model's capacity to accurately delineate relevant areas in the mpMRI images of prostate cancer. The algorithm has been tested on the PROMISE12 dataset [44].

- **nnU-Net**

  The research project focused on the application of the nnU-Net model to segment mpMRI prostate scans, aiming to accurately identify regions associated with prostate cancer. The nnU-Net model had undergone pretraining to enhance its performance prior to the study. The main objective was to achieve precise delineation of cancerous areas in the images. Notably, the nnU-Net model outperformed other models utilized in the research, demonstrating the highest level of accuracy. With a remarkable dice coefficient of 0.951, the model exhibited a strong agreement between its predicted segmentations and the ground truth. These findings underscore the efficacy of leveraging pretrained models and emphasize the immense potential of the nnU-Net architecture in advancing prostate cancer segmentation tasks. The findings of this study carry substantial implications for improving the detection and management of prostate cancer by leveraging the benefits of mpMRI scans. The algorithm has been tested on the PROMISE12 dataset ProstateX [45].

  The performance of four distinct automated segmentation models, namely U-Net, V-net, 3D U-Net and nnU-Net, was evaluated using the dataset. The performance of implementation is accessed with parameter Dice-coefficient, number of epoch and processing time. Table 7.1 shows the comparative analysis of automated segmentation using Dice-coefficient and nnU-Net outperforms other techniques.

Table 7.1. Comparative analysis of automated segmentation using Dice-coefficient.

Model	U-Net	3D U-Net	V-net	nnU-Net
Dice-coefficient	0.913	0.926	0.949	0.951
No. of epoch	150	15 000	100	1000
Time to process (hours)	2.58	9.25	20	X[1]

Notes[1] Pretrained model used.

7.5.2 Classification results

- **ResNet50**

  ResNet50 was employed as one of the models to classify mpMRI scans of the prostate into malignant (cancerous) and benign (non-cancerous) categories. In this study, the dataset comprising of 2160 images was utilized and subsequently divided into separate training and test sets with a batch size of 16, ResNet50 underwent training and evaluation. The final results indicated an accuracy of 0.828. The precision and recall for classifying benign samples were 0.54 and 0.57, respectively, the outcome led to an F1-score of 0.70. For the malignant class, ResNet50 achieved a recall of 0.61, a precision of 0.41 and an F1-score of 0.48. These findings highlight the model's ability to effectively differentiate between cancerous and non-cancerous mpMRI scans, although its performance varied between the two classes (figure 7.11).

- **MobileNet**

  MobileNet, another model utilized in the study, was employed to classify mpMRI scans into malignant and benign categories. Similar to ResNet50, the data set underwent a partition into training and test sets, while the model underwent training using a batch size of 16. The final evaluation revealed an accuracy of 0.818. MobileNet achieved a precision of 0.58 and a recall of 0.60 for classifying benign samples, the outcome led to an F1-score of 0.66. For the malignant class, the model achieved a precision of 0.52, a recall of 0.62, and an F1-score of 0.56. These results demonstrate MobileNet's ability to accurately classify cancerous and non-cancerous mpMRI scans, indicating promising performance for this classification task (figure 7.12).

- **DenseNet201**

  DenseNet201, another model used in the research, aimed to classify mpMRI scans into malignant and benign categories. Trained with a batch

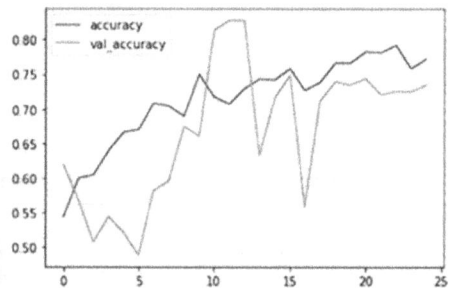

(a)

(b)

Figure 7.11. Performance analysis (a) ResNet-50 model parameters (b) performance of ResNet50.

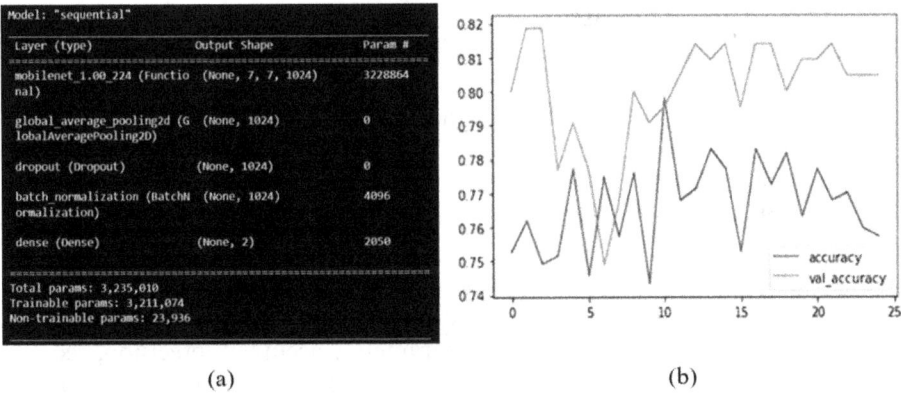

(a)   (b)

Figure 7.12. Performance analysis (a) MobileNet model parameters (b) performance of MobileNet.

- **DenseNet201**

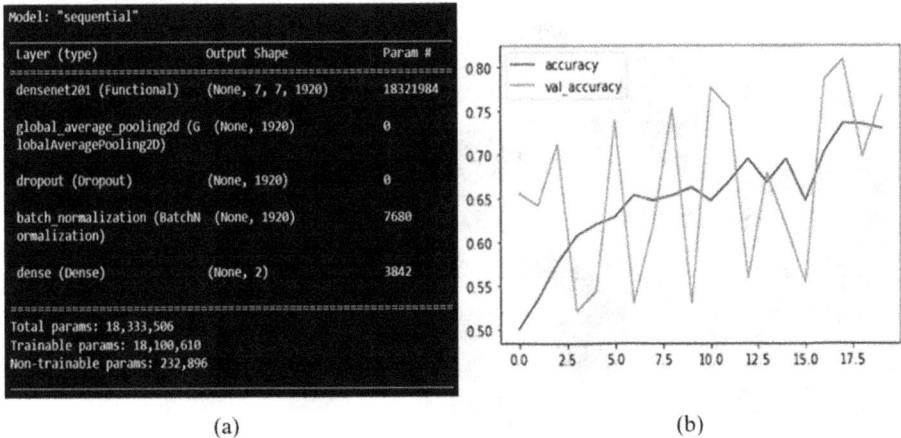

(a)   (b)

Figure 7.13. Performance analysis (a) DenseNet201 model parameters (b) performance of DenseNet201MobileNet.

size of 5 using the dataset of 2160 images, DenseNet201 achieved an accuracy of 0.809. The precision and recall for classifying benign samples were 0.58 and 0.62, respectively, resulting in an F1-score of 0.68. For the malignant class, the precision of the model was measured at 0.51, while its recall stood at 0.63, resulting in an F1-score of 0.56. These findings highlight DenseNet201's potential in accurately distinguishing between cancerous and non-cancerous mpMRI scans, although its performance varied slightly between the two classes (figure 7.13).

- **InceptionV3**

InceptionV3, the final model used in the study, was employed to classify mpMRI scans into malignant and benign categories. Trained using a batch size of 16 and the dataset of 2160 images, an accuracy of 0.819 was attained. The recall and precision for classifying benign samples were 0.59 and 0.57, respectively, this led to an F1-score of 0.62. For the malignant class, the precision, recall, and F1-score of the model were recorded as 0.54, 0.59, and 0.56, respectively. These results demonstrate InceptionV3's potential in accurately classifying cancerous and non-cancerous mpMRI scans, with relatively consistent performance for both classes (figure 7.14).

Table 7.2 shows the comparative analysis of four classification models used to classify the mpMRI as benign and malignant. The classification performance of algorithms accessed with parameters accuracy, precision, recall and F1-score, CNN ResNet-50 classification model outperforms other models.

- **InceptionV3**

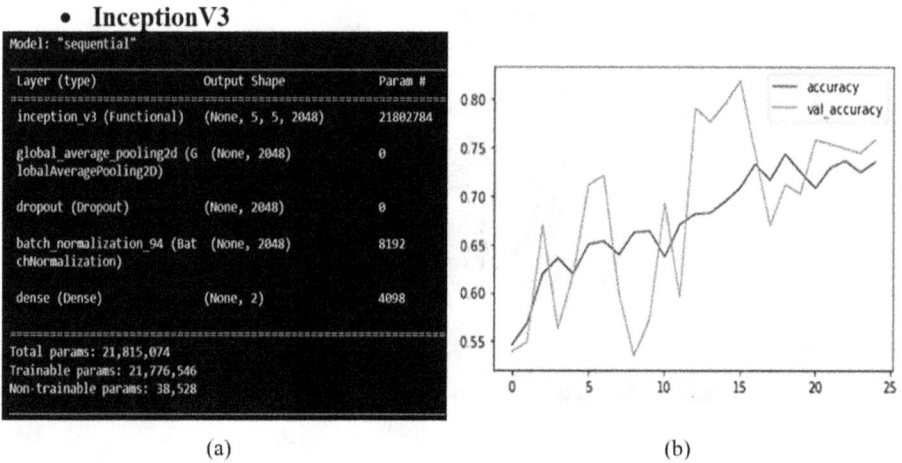

(a)   (b)

Figure 7.14. Performance analysis (a) InceptionNetv3 model parameters (b) performance of InceptionNetv3.

Table 7.2. Comparative analysis of classification model.

		Model performance					
		Precision		Recall		F1-score	
Model name	Accuracy	Benign	Malignant	Benign	Malignant	Benign	Malignant
ResNet-50	0.828	0.54	0.57	0.70	0.41	0.61	0.48
MobileNet	0.818	0.58	0.60	0.66	0.52	0.62	0.56
DenseNet201	0.809	0.58	0.62	0.68	0.51	0.63	0.56
InceptionV3	0.819	0.57	0.59	0.62	0.54	0.59	0.56

7.6 Conclusion

The comparative analysis of deep neural network architectures for prostate region segmentation and classification in multi-parametric magnetic resonance imaging (mpMRI) provides significant insights into their performance and potential applications in prostate cancer diagnosis and treatment. The evaluated segmentation models, including U-Net, V-Net, 3D U-Net, and nnUNet showcased varying levels of accuracy in delineating the peripheral zone and central gland regions of the prostate. Similarly, the classification models, ResNet50, MobileNet, DenseNet201, and InceptionNetV3, exhibited comparable abilities in distinguishing between benign and malignant tumors. These findings highlight the significance of deep learning techniques in improving the precision and effectiveness of prostate cancer analysis using mpMRI. Further advancements in this field possesses the capability to enhance diagnosis accuracy, provide assistance in developing treatment plans and ultimately contribute to improved patient outcomes. Continued research and development in deep learning methodologies are crucial for unlocking the complete potentiality of these models in prostate cancer management.

7.6.1 Scope for future work

The future of artificial intelligence in prostate cancer analysis through segmentation as well as classification holds great promise. Advancements in AI technologies, such as deep learning and convolutional neural networks, can improve accuracy and efficiency in detecting prostate cancer, enabling early diagnosis and intervention for better patient outcomes. Integrating AI with multimodal data and refining existing models like nnU-Net and ResNet-50 can further enhance performance. Continued research aims to revolutionize prostate cancer diagnosis and treatment, leading to improved precision and personalized care.

References

[1] Hassan M R, Islam M F, Uddin M Z, Ghoshal G, Hassan M M, Huda S and Fortino G 2022 Prostate cancer classification from ultrasound and MRI images using deep learning based explainable artificial intelligence *Future Gener. Comput. Sys.* **127** 462–72

[2] Mirasol I V O, Abu P A R and Reyes R S 2022 Construction of a repeatable framework for prostate cancer lesion binary semantic segmentation using convolutional neural networks *Int. J. Adv. Comput. Sci. Appl.* **13** 1

[3] Fernandez-Quilez A 2022 Deep learning for an improved diagnostic pathway of prostate cancer in a small multi-parametric magnetic resonance data regime *Doctoral Thesis* University of Stavanger

[4] Zhao L *et al* 2022 A deep learning approach for predicting clinically significant prostate cancer: a retrospective, multicentre study *Eur. J. Nucl. Med. Mol. Imaging* **50** 727–41

[5] Daneshvar M *et al* 2022 PD40-05 Deep learning based assessment of prostate lesion dynamics on multiparametric MRI during active surveillance *J. Urol.* **207** e685

[6] Zaridis D, Mylona E, Tachos N, Marias K, Tsiknakis M and Fotiadis D I 2022 A smart cropping pipeline to improve prostate's peripheral zone segmentation on MRI using deep learning *EAI Endorsed Trans. Bioeng. Bioinform.* **1** e3

[7] Chun C W, Yousir N T, Abdulameer S M, Mostafa S A and Hezam A A 2022 Deep learning approach for predicting prostate cancer from MRI images *J. Soft Comput. Data Min.* **3** 1–9

[8] Mehralivand S *et al* 2022 A cascaded deep learning–based artificial intelligence algorithm for automated lesion detection and classification on biparametric prostate magnetic resonance imaging *Acad. Radiol.* **29** 1159–68

[9] Ye L Y, Miao X Y, Cai W S and Xu W J 2022 Medical image diagnosis of prostate tumor based on PSP-Net+ VGG16 deep learning network *Comput. Methods Programs Biomed.* **221** 106770

[10] Hosseinzadeh M, Saha A, Brand P, Slootweg I, de Rooij M and Huisman H 2022 Deep learning–assisted prostate cancer detection on bi-parametric MRI: minimum training data size requirements and effect of prior knowledge *Eur. Radiol.* 1–11

[11] Hu L, Zhou D W, Guo X Y, Xu W H, Wei L M and Zhao J G 2022 Adversarial training for prostate cancer classification using magnetic resonance imaging *Quant. Imaging Med. Surg.* **12** 3276

[12] Koonamparampath M, Shah R, Sundvesha M and Ugale M 2022 A review on prostate cancer detection using CNN *Int. J. Res. Appl. Sci. Eng. Technol. (IJRASET)* **10** 948–53

[13] Roest C, Kwee T C, Saha A, Fütterer J J, Yakar D and Huisman H 2023 AI-assisted biparametric MRI surveillance of prostate cancer: feasibility study *Eur. Radiol.* **33** 89–96

[14] Bleker J, Kwee T C, Rouw D, Roest C, Borstlap J, de Jong I J, Dierckx R A, Huisman H and Yakar D 2022 A deep learning masked segmentation alternative to manual segmentation in biparametric MRI prostate cancer radiomics *Eur. Radiol.* **32** 6526–35

[15] Pellicer-Valero O J, Marenco Jimenez J L, Gonzalez-Perez V, Casanova Ramon-Borja J L, Martin Garcia I, Barrios Benito M, Pelechano Gomez P, Rubio-Briones J, Rupérez M J and Martín-Guerrero J D 2022 Deep learning for fully automatic detection, segmentation, and Gleason grade estimation of prostate cancer in multiparametric magnetic resonance images *Sci. Rep.* **12** 2975

[16] Salvi M, De Santi B, Pop B, Bosco M, Giannini V, Regge D, Molinari F and Meiburger K M 2022 Integration of deep learning and active shape models for more accurate prostate segmentation in 3D MR images *J. Imaging* **8** 133

[17] Künzel L A, Nachbar M, Hagmüller M, Gani C, Boeke S, Wegener D, Paulsen F, Zips D and Thorwarth D 2022 Clinical evaluation of autonomous, unsupervised planning integrated in MR-guided radiotherapy for prostate cancer *Radiother. Oncol.* **168** 229–33

[18] Fan X *et al* 2022 Multiparametric MRI and machine learning based radiomic models for preoperative prediction of multiple biological characteristics in prostate cancer *Front. Oncol.* **12** 839621

[19] Quihui-Rubio P C, Ochoa-Ruiz G, Gonzalez-Mendoza M, Rodriguez-Hernandez G and Mata C 2022 Comparison of automatic prostate zones segmentation models in MRI images using U-net-like architectures *Mexican Int. Conf. on Artificial Intelligence* (Cham: Springer Nature) pp 282–96

[20] Jing G, Xing P, Li Z, Ma X, Lu H, Shao C, Lu Y, Lu J and Shen F 2022 Prediction of clinically significant prostate cancer with a multimodal MRI-based radiomics nomogram *Front. Oncol.* **12** 918830

[21] Asavareongchai N, Banerjee I, Tea-Makorn P P and Phulsuksombati M 2017 *Multiparametric MR Image Analysis for Prostate Cancer Assessment with Convolutional Neural Networks*

[22] Adams L C *et al* 2022 Prostate158-An expert-annotated 3T MRI dataset and algorithm for prostate cancer detection *Comput. Biol. Med.* **148** 105817

[23] Li B, Oka R, Xuan P, Yoshimura Y and Nakaguchi T 2022 Robust multi-modal prostate cancer classification via feature autoencoder and dual attention *Inform. Med. Unlocked* **30** 100923

[24] Tyagi S, Tyagi N, Choudhury A, Gupta G, Zahra M M A and Rahin S A 2022 Identification and classification of prostate cancer identification and classification based on improved convolution neural network *Bio. Med. Res. Int.* 2022

[25] Twilt J J, van Leeuwen K G, Huisman H J, Fütterer J J and de Rooij M 2021 Artificial intelligence based algorithms for prostate cancer classification and detection on magnetic resonance imaging: a narrative review *Diagnostics* **11** 959

[26] Cipollari S, Pecoraro M, Forookhi A, Laschena L, Bicchetti M, Messina E, Lucciola S, Catalano C and Panebianco V 2022 Biparametric prostate MRI: impact of a deep learningbased software and of quantitative ADC values on the inter-reader agreement of experienced and inexperienced readers *Radiol. Med.* **127** 1245–53

[27] Hu B *et al* 2021 Classification of prostate transitional zone cancer and hyperplasia using deep transfer learning from disease-related images *Cureus* 13

[28] Collado C N 2022 Comprehensive study of good model training for prostate segmentation in volumetric MRI *arXiv preprint* arXiv:2208.13671

[29] Yin H and Buduma N 2022 Prostate lesion detection and salient feature assessment using zone-based classifiers *arXiv preprint* arXiv:2208.11522

[30] van Sloun R J, Wildeboer R R, Mannaerts C K, Postema A W, Gayet M, Beerlage H P, Salomon G, Wijkstra H and Mischi M 2021 Deep learning for real-time, automatic, and scanner-adapted prostate (zone) segmentation of transrectal ultrasound, for example, magnetic resonance imaging–transrectal ultrasound fusion prostate biopsy *Eur. Urol. Focus.* **7** 78–85

[31] Litjens G, Toth R, Van De Ven W, Hoeks C, Kerkstra S, Van Ginneken B, Vincent G *et al* 2014 Evaluation of prostate segmentation algorithms for MRI: the PROMISE12 challenge *Med. Image Anal.* **18** 359–73 ('Publicly available datasets': PROMISE12: Data from the MICCAI Grand Challenge: Prostate MR Image Segmentation 2012 | Zenodo)

[32] Litjens G, Debats O, Barentsz J, Karssemeijer N and Huisman H 2017 ProstateX challenge data The Cancer Imaging Archive 2017.MURS5CL ('Publicly available datasets': SPIE-AAPM-NCI PROSTATEx Challenges (PROSTATEx)—The Cancer Imaging Archive (TCIA) Public Access—Cancer Imaging Archive Wiki)

[33] IA92 n.d. PROMISE12_OptUNet GitHub https://github.com/IA92/PROMISE12_OptUNet/blob/master/PROMISE12_Notebook.ipyn (Accessed: March 23, 2023)

[34] jancio n.d. 3D-U-Net-Prostate-Segmentation [GitHub Repository] https://github.com/jancio/3D-U-Net-Prostate-Segmentation (accessed 23 March 2023)

[35] junqiangchen n.d. VNet [GitHub Repository] https://github.com/junqiangchen/VNet (accessed 23 March 2023)

[36] MIC-DKFZ n.d. nnUNet: nnunetv1 branch [GitHub Repository] https://github.com/MICDKFZ/nnUNet/tree/nnunetv1 (accessed 23 March 2023)

[37] AthiraNirmal n.d. Brain Cancer_Detection.ipynb [GitHub Repository] https://github.com/AthiraNirmal/Medical-Image-Classification-/blob/main/Brain%20Cancer_Detection.ipynb (accessed 23 March 2023)

[38] Ronneberger O, Fischer P and Brox T 2015 U-net: convolutional networks for biomedical image segmentation *Medical Image Computing and Computer-Assisted Intervention–MICCAI 2015: 18th Int. Conf., Munich, Germany* (Cham: Springer International) pp 234–41

[39] Çiçek Ö, Abdulkadir A, Lienkamp S S, Brox T and Ronneberger O 2016 3D U-Net: learning dense volumetric segmentation from sparse annotation *Medical Image Computing and Computer-Assisted Intervention–MICCAI 2016: 19th Int. Conf., Athens, Greece* (Cham: Springer International) pp 424–32

[40] Milletari F, Navab N and Ahmadi S A 2016 V-Net: fully convolutional neural networks for volumetric medical image segmentation *2016 4th Int. Conf. on 3D Vision (3DV)* (Piscataway, NJ: IEEE) pp 565–71

[41] Isensee F, Jaeger P F, Kohl S A, Petersen J and Maier-Hein K H 2021 nnU-Net: a selfconfiguring method for deep learning-based biomedical image segmentation *Nat. Methods* **18** 203–11

[42] He K, Zhang X, Ren S and Sun J 2016 Deep residual learning for image recognition *Proc. of the IEEE Conf. on Computer Vision and Pattern Recognition* pp 770–8

[43] Howard A G, Zhu M, Chen B, Kalenichenko D, Wang W, Weyand T, Andreetto M and Adam H 2017 Mobilenets: efficient convolutional neural networks for mobile vision applications *arXiv preprint* arXiv:1704.04861

[44] Huang G, Liu Z, Van Der Maaten L and Weinberger K Q 2017 Densely connected convolutional networks *Proc. of the IEEE Conf. on Computer Vision and Pattern Recognition* pp 4700–8

[45] Szegedy C, Liu W, Jia Y, Sermanet P, Reed S, Anguelov D, Erhan D, Vanhoucke V and Rabinovich A 2015 Going deeper with convolutions *Proc. IEEE Conf. on Computer Vision and Pattern Recognition* pp 1–9

IOP Publishing

Image Processing with Python
A practical approach
Irshad Ahmad Ansari and Varun Bajaj

Chapter 8

Optical character recognition: transforming images into text

Nikhil Kushwaha, Om Asati and Mainak Sadhya

Optical character recognition (OCR) is a technology that enables the full recognition of the alphanumeric printed characters. It scans the text image then interprets the printed characters and converts it into a corresponding editable text document. This chapter describes the working of the OCR to extract the text content of the image and the implementation of the convolutional neural network to extract the features of the image. ResNet (Residual Network) is a deep learning architecture that addresses the problem of degradation and vanishing gradients in very deep neural networks. Optical character recognition can also be achieved using different programming languages like Java, C++, C#, JavaScript, Ruby, MATLAB, etc., but because of less complexity, less syntax and easy to use functionalities of a bunch of diverse libraries, Python is the most used language in such fields as image processing and video processing.

8.1 Introduction

In today's digital age, the need to detect alphanumeric content (text) in images to convert it into machine-readable format has become increasingly critical [1]. Optical character recognition (OCR) technology has emerged as a powerful solution to address this demand. OCR plays a very pivotal role in automating the extraction of textual information from various sources, such as scanned documents, images and photographs. It enables the transformation of visual data into editable and searchable text. [2] involves the OCR system training a model using machine learning algorithms. By feeding the model to a large dataset of labeled images containing text, it learns to recognize different characters and patterns. Over time, the model becomes more accurate [3]. The aim of this technology is to show that OCR can detect the alphanumeric characters present in the number plate of any vehicle. The integration of OCR with CCTV cameras brings several benefits to the

doi:10.1088/978-0-7503-5924-5ch8

traffic management. It enhances the efficiency by automating the identification process, reducing the need of manual intervention [3]. This enables quick identification of vehicles, allowing authorities to promptly address traffic violations and maintain road safety. By digitalizing this process, it also reduces paperwork and simplifies record keeping, leading to streamlined operations within the RTO [2].

8.2 Implementation

8.2.1 OCR Model Structure (flow diagram)

Figure 8.1 shows the developed OCR model.

8.2.2 Explanation

8.2.2.1 Data collection

As shown in the diagram, the very first step is to collect the dataset from various sources like Kaggle([4] A–Z hand written dataset, [5] 0–9 dataset, [6] a–z dataset), [7] University of Surrey (0–9 dataset, A–Z dataset, a–z dataset), MNIST dataset and then combining all the datasets into a Numpy array with 62 labels (0–9, a–z, A–Z) [3].

8.2.2.2 Data preprocessing

In this section, it can be seen that the change in dimension of the images can lead to reduction in noise as it is required to change the shape of each image to 28×28 using a cv2.resize function [3]. Reducing the size of the image increases the computational

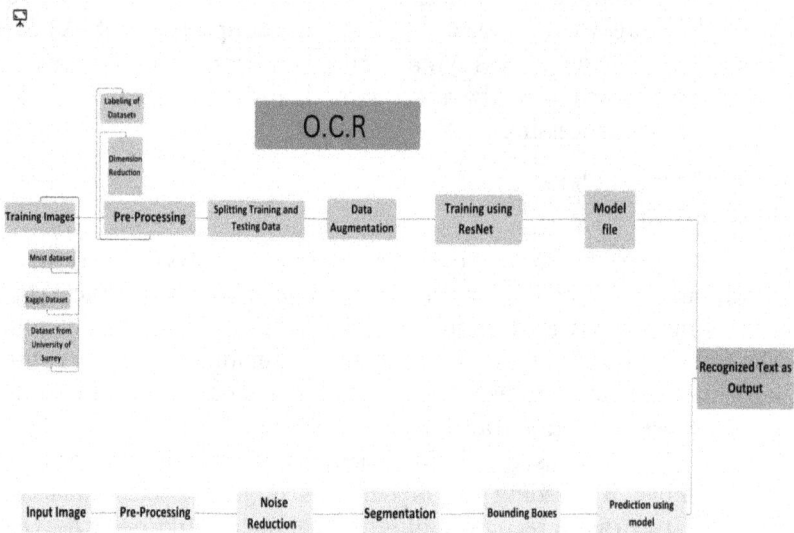

Figure 8.1. Schematic work flow diagram of OCR model.

efficiency for processing as the smaller images occupy less memory and allow the OCR model to handle the large amount of data [1].

8.2.2.2.1 Python code that can resize the image to 28·28

```python
import cv2
# OpenCv(cv2) is a popular computer vision library. It is used in image and video
processing. import numpy as np
#Numpy is for handling of multi-dimensional arrays and matrices

def convert_to_28x28(i_p): #i_p stands for image path
    # Read and convert the image to grayscale using OpenCV's built-in function
    image = cv2.cvtColor(cv2.imread(i_p), cv2.COLOR_BGR2GRAY)

    # Conversion of each image to the shape of (28, 28, 1)
    wdth, hieght = imge.shape

    if hieght > 28 or wdth > 28:
        # Resize the image using OpenCV's built-in functions
        (trgt_hieght, trgt_wdth) = imge.shape
        x = max(0, 28 - trgt_wdth)
        y = max(0, 28 - trgt_hieght)
        dlta_x = int(x / 2.0)
        dlta_y = int(y / 2.0)
        cmb = cv2.copyMakeBorder
        # Add zero-pixel borders to the image to match the target
        size imge = cmb(imge, tp=dlta_y,botm=dlta_y,
                    lft=dlta_x, rht=dlta_x,
                    bdrTyp=cv2.BORDER_CONSTANT, value=(0, 0, 0))

        # Resize the image to the target size of (28, 28)
        imge = cv2.resize(imge, (28, 28))

    wdth, hieght = imge.shape

    if wdth < 28:
```

```
# Concatenate the image with additional rows of zeros using numpy's built-in functions
addng_zros = np.ones((28 - wdth, hieght)) * 255
imge = np.concatenate((imge, addng_zros))

if hieght < 28:
    # Concatenate the image with additional columns of zeros using numpy's built-in functions
    addng_zros = np.ones((28, 28 - hieght)) * 255
    imge = np.concatenate((imge, addng_zros), axis=1)

# Normalize each image
return imge
```

Then combining the MNIST dataset and the Kaggle dataset (handwritten A-Z, 0–9, a–z) and dataset from University of Surrey (A–Z, 0–9, a–z).

First it is needed to load the dataset from Kaggle, which is in .csv format, and also reshape it to 28×28 [1].

8.2.2.2.2 *Python code to load and reshape the dataset*

```
import numpy as np
#Numpy is for handling of multi-dimensional arrays and matrices
import pandas as pd
# Pandas is library for data manipulation and data analysis


def load_custom_dataset(dtaset_pth):
    dta = []
    lbls = []
    df = pd.read_csv(dtaset_pth)

    # Iterate over each row in the
    dataset for _, row in df.iterrows():
        cstm_lbl = int(row[0])
        cstm_imge = np.array([int(x) for x in row[1:]],
        dtype="uint8") cstm_imge = cstm_imge.reshape((28, 28))
        dta.append(cstm_imge)
        lbls.append(cstm_lbl)

    cstm_dta = np.array(dta, dtype='float32')
    cstm_lbls = np.array(lbls, dtype="int")
    return (cstm_dta, cstm_lbls)
```

After loading this dataset it is required to combine all three datasets and save it as a .npy file. Now, we use this combined resized file (28×28) to train the model. It is required to create 62 labels as : labels=**'0123456789ABCDEFGHIJKLMNOP QRSTUVWXYZabcdefghijklmnopqrstuv wxyz'**

These labels are needed to predict the particular character by comparing input images with it.

To combine all three datasets using Numpy, combine the final dataset file as a vertical stack using **np.vstack([mnist_data, data, azData])**

Note: mnist_data is from MNIST, and consists of 60 000 trainData and 10 000 testData, data is the dataset from University of Surrey and azData is the dataset from Kaggle

8.2.2.2.3 Python code to combine all the datasets and save it as a .npy file

```python
from keras.datasets import mnist
# Built-in dataset of handwritten digits for training and testing machine learning models
from keras.preprocessing.image import ImageDataGenerator
# Image preprocessing utilities module for deep learning models
import numpy as np
#Numpy is for numerical computing and supports for large multi-dimensional arrays andmatrice
import datas as dt
import datasetpath as dsp
import cv2
# OpenCV library for computer vision tasks and image processing
from sklearn.preprocessing import LabelBinarizer
# It is the library for transforming categorical labels into binary vectors
from sklearn.model_selection import train_test_split
# Function for splitting datasets into training and testing subsets

((trnDta, trnLbls), (tstDta, tstLbls)) = mnist.load_data()
(azDta, azLbls) = dsp.load_custom_dataset('/Users/nikhilkushwaha/ocr/A_Z Handwritten Data.csv')
# print(azData.shape)

cmbnd_dta = np.vstack([trnDta, tstDta])
cmbnd_lbls = np.hstack([trnLbls, tstLbls])


# print(trnDta.shape)
# print(cmbnd_dta[0].shape)
# print(dt.lbls.shape)


azLbls+=10
# combining all the dataset
lbls1=np.hstack([cmbnd_lbls,dt.labels,azLbls])
dta1=np.vstack([cmbnd_dta,dt.data,azDta])
# print(lbls1.shape)
# print(dta1.shape)

# print(labels1.shape)

# print(data1.shape)

# saving the combined data
np.save('/Users/nikhilkushwaha/ocr/combined_data.npy', dta1)
np.save('/Users/nikhilkushwaha/ocr/combine_labels.npy', lbls1)


# loading the combined data and labels


loded_dta = np.load('/Users/nikhilkushwaha/ocr/combined_data.npy')
loded_lbls = np.load('/Users/nikhilkushwaha/ocr/combine_labels.npy')

cv2.imshow("image" , loded_dta[-50000])
# cv2.waitKey(0)
# cv2.destroyAllWindows()

tmp_image = cv2.resize(loded_dta[0], (32, 32))
```

```
# print(tmp_image.shape)

dta = [cv2.resize(image, (32, 32)) for image in
loded_dta] dta = np.array(dta, dtype="float32")

# print(data.shape)

dta2 = np.expand_dims(dta, axis=-1)

# print(data2.shape)
```

8.2.2.3 Label encoding

As machine learning algorithms work on the numerical data, it cannot directly handle the categorical labels. So it is needed to encode categorical labels into binary vectors, which makes easy to train the OCR model [1].

8.2.2.3.1 Python code to encoding labels

```
lbl_bin = LabelBinarizer()
bnrized_lbls = lbl_bin.fit_transform(loded_lbls)
class_totals = bnrized_lbls.sum(axis=0)
# print(binarized_labels[0])
```

8.2.2.4. Loading class weights

62 labels are created, so it is needed to load 62 classweights, which is done by

```
classWeight     =     labels.sum(axis=0).max()/labels.sum
(axis=0)
```

8.2.2.4.1 Python code to load class weights

```
# classweight
class_totals = bnrized_lbls.sum(axis=0)
class_weight = {}
# loop over all classes and calculate the class
weight for i in range(0, len(class_totals)):
        class_weight[i] = class_totals.max() / class_totals[i]

# print(class_weight)
```

8.2.2.5 Splitting the data into 75% training and 25% testing dataset

It is needed to split the data into some ratio like 75% as a training dataset, which is referred as train_data_X, train_data_Y, and 25% as a test dataset, which is referred as test_data_X and test_data_Y, so that the model will predict what we required, when any unseen data is given at the time of evaluation. It is to not let the model be underfit or overfit.

8.2.2.5.1 Python code for train-test split

```
from sklearn.model_selection import train_test_split # Function for splitting datasets into training and
testing subsets
(trn_dta_X, tst_dta_X, trn_dta_Y, tst_dta_Y) = train_test_split(dta2,
        bnrized_lbls, test_size=0.25, stratify=bnrized_lbls, random_state=42)

# print(trn_dta_X.shape)
# print(trn_dta_X.shape)
```

8.2.2.6 Building of residual network

This model is using Residual Network (ResNet) to train as it is based on deep convolution neural network architecture. [2] ResNet's deep layers capture the low-level features of an image like edges and textures, while higher-level layers of the ResNet capture more abstract features like character shapes and arrangements. [3] ResNet can directly be used by importing it as `ResNet50` from `keras.applications`, but it can also be implemented by creating ResNet Module as shown below.

8.2.2.6.1 Python code for building residual network (ResNet)

First we need to import all the required libraries from the `keras.layers` (which is a high-level API of the Tensorflow).

```
from keras.models import Model
# Model class that represents a deep learning model
from keras.layers.convolutional import ZeroPadding2D
# Zero padding layer that pads the input with zeros
from keras.layers import BatchNormalization
# It is a technique used to normalize the activations of a previous layer
from keras.layers.core import Activation
from keras.layers.convolutional import Conv2D
# This module of keras.layers is performs convolutional operations on 2D inputs
from keras.regularizers import l2
# L2 regularization that applies a penalty on the layer's
weights from keras.layers import Flatten
# It Flattens the input to a 1D array
from keras.layers.convolutional import AveragePooling2D
# This layer performs downsampling using average values
from keras.layers import Input
# Input layer for defining the input
shape from keras.layers import add
# Element-wise addition layer
from keras.layers.convolutional import MaxPooling2D
# Max pooling layer that performs downsampling using maximum
values from keras import backend as K
# Backend module for Keras that provides low-level operations and functions
from keras.layers.core import Dense
# Fully connected dense layer that applies a specified number of output units
```

Then we define a method under **class ResNet**: the **residual_module** method, which has parameters as follows:

1. **dta**—This variable is for the input to the method.
2. **nm_fltrs**—This variable is to take care of the number of filters (output channels) for the convolutional layer.
3. **strde**—This variable is for the stride in the convolutional layers.
4. **chnnl_Dim**: The dimension's index of the channel in the input data.
5. **rdce**: A Boolean flag indicating whether the module is used in reducing the spatial size.
6. **rgulariztion**: The regularization parameter for the convolutional layers.
7. **bn_epsln and bn_momntm**: These are the batch normalization parameters.

    Batch normalisation is a technique to normalise the inputs of a neural network layer by transforming them to having mean = 0 and variance = 0.

    Inside this method, the residual module is constructed with three blocks.
8. The first block consists of three layers: batch normalization layer, activation layer (RELU) and a 1×1 convo layer.
9. The second block consists three layers: batch normalization, activation and a 3×3 convo layer.
10. The third layer is another batch normalisation, activation and 1×1 convo layer.
11. If the 'reduce' flag is 1 (set), a 1×1 convolutional layer is applied to the shortcut connection to reduce its spatial size.
12. Finally, output of the residual module is obtained by adding the convolutional output and the shortcut connection.

    It is required to define the **build method,** which constructs the **ResNet architecture**. It takes parameters such as the **input dimensions (width, height, depth), the number of the classes, the number of stages, and the number of filters per stage.**

```
class RsNt:
    @staticmethod
    def rsdual_mod(dta, nm_fltrs, strde, chnnl_dim, rdce=False,
                rgularztion=0.0001, bn_epsln=2e-5, bn_momntm=0.9):

        # The shortcut branch of the RsNt module is init as the input (identity) data
        shrtct = dta

        # The 1st block of the RsNt module consisting of 1x1 convo's
        btchnrm1 = BatchNormalization(axis=chnnl_dim, epsilon=bn_epsln,
                        momentum=bn_momntm)(dta)
        actvtion1 = Activation("relu")(btchnrm1)
        conv1 = Conv2D(int(nm_fltrs * 0.25), (1, 1), use_bias=False,
                kernel_regularizer=l2(rgularztion))(actvtion1)

        # The 2nd block of the ResNet module consisting of 3x3 convo's
        btchnrm2 = BatchNormalization(axis=chnnl_dim, epsilon=bn_epsln,
                        momentum=bn_momntm)(conv1)
        actvtion2 = Activation("relu")(btchnrm2)
        conv2 = Conv2D(int(nm_fltrs * 0.25), (3, 3), strides=strde,
                padding="same", use_bias=False,
                kernel_regularizer=l2(rgularztion))(actvtion2)

        # The 3rd block of the RsNt module consisting of another set of 1x1 convo's
        batchnorm3 = BatchNormalization(axis=chnnl_dim, epsilon=bn_epsln,
                        momentum=bn_momntm)(conv2)
        actvtion3 = Activation("relu")(batchnorm3)
        conv3 = Conv2D(nm_fltrs, (1, 1), use_bias=False,
                kernel_regularizer=l2(rgularztion))(actvtion3)

        # If we need to reduce the spat size, apply a convo layer to the
        shortcut if rdce:
            shrtct = Conv2D(nm_fltrs, (1, 1), strides=strde,
                    use_bias=False, kernel_regularizer=l2(rgularztion))(actvtion1)

        # Add together the shortcut and the final
        convo x = add([conv3, shrtct])

        # Ret the add as the output of the RsNt module
        return x
```

13. The input shape is defined based on the data format (channel last or channel first). Then it is needed to apply the batch normalisation to the inputs [8].

```
@staticmethod
  def build(wdth, hieght, dpth, nm_clsses, stgs, fltrs,
        rgularztion=0.0001, bn_epsln=2e-5, bn_momntm=0.9):
      # Initialize the input shape to be "channels last" and the channels dimension itself
      inpt_shpe = (hieght, wdth, dpth)
      chnnl_dim = -1

      # If we are using "channels first", update the input shape and channels dimension
      if K.imge_dta_format() == "channels_first":
          inpt_shpe = (dpth, hieght, wdth)
          chnnl_dim = 1

      # Set the input and apply batch normalization
      inpts = Input(shape=inpt_shpe)
      x = BatchNormalization(axis=chnnl_dim, epsilon=bn_epsln,
                  momentum=bn_momntm)(inpts)

      # Loop over the number of
      stages for i in range(0, len(stgs)):
          # First we need to Init the stride , then applying a rsdual_mod
          # it is used to reduce the spat size of the input volume
          stride = (1, 1) if i == 0 else (2, 2)
          x = RsNt.rsdual_mod(x, fltrs[i + 1], stride,
                      chnnl_dim, reduce=True, bn_epsilon=bn_epsln, bn_momentum=bn_momntm)

          # Loop over the number of layers in the
          stage for j in range(0, stgs[i] - 1):

              # Aply a RsNt mdule

              x = RsNt.rsdual_mod(x, fltrs[i + 1],
                          (1, 1), chnnl_dim, bn_epsilon=bn_epsln, bn_momentum=bn_momntm)

      # Apply batch normalization => activation => pooling
      x = BatchNormalization(axis=chnnl_dim, epsln=bn_epsln,
                  momentum=bn_momntm)(x)
      x = Activation("relu")(x)
      x = AveragePooling2D((8, 8))(x)

      # Softmax classifier
      x = Flatten()(x)
      x = Dense(nm_clsses, kernel_regularizer=l2(rgularztion))(x)
      x = Activation("softmax")(x)

      # Create the model
      model = Model(inpts, x, name="resnet")

      # Return the constructed network
      architecture
      return model
```

14. Then depending on the dataset, the initial convolutional layer is added with the specified filter size.
15. Then for each stage, a series residual is added and the stride of the first module in each stage is set to (2, 2) to reduce the spatial size.
16. Within each stage, multiple residual modules are added.
17. Batch normalization and the activation are applied after the last stage.
18. Average pooling is applied to reduce the spatial dimensions.

19. The output is flattened and passed through a fully connected layer with the specified number of classes.
20. Finally, a softmax activation is applied to reduce the spatial to produce the class probabilities.
21. Then, creating and returning the constructed network architecture.

8.2.2.7 *Compiling and training of model*

Now, the ResNet module is defined so we need to compile the model so that it undergoes training. The model building is done by using **RsNt.build** by choosing these parameters for build method (**wdth** $= 32$, **hieght** $= 32$, **dpth**$=$ 1, **clsses** $=$ len(**lbl_bin. classes_**), **stgs** $= (3, 3, 3)$, **fltrs** $= (64, 64,$ 128, 256)$, **reg** $= 0.0005)$

8.2.2.7.1 *Python code for compiling and training of the model*

```python
# Import the 'resnet' module as 'rsnt' for ResNet
architecture import resnet as rsnt

# Import the 'Adam' optimizer from the 'keras.optimizers' library
from keras.optimizers import Adam

# Import the 'mnistdataset' module as 'mndt' for the MNIST
dataset import mnistdataset as mndt

# Import the 'numpy' library as 'np' for numerical
computing import numpy as np

# Import the 'os' library for operating system related functionalities
import os

# Import the 'tensorflow' library as 'tf' for machine learning and deep learning
import tensorflow as tf

# Import the 'load_model' function from the 'keras.models' library for loading pre-trained
models from keras.models import load_model

# Set the epochs counts
nm_epchs = 1

# Set the init learning rate
lrning_rte = 0.01

# Set the batch size
btch_sze = 256
```

```
print("[INFO] Compiling the ocr model...")
optmzr = Adam(learning_rate=lrning_rte)

# Build the ResNet model
model = rsnt.RsNt.build(32, 32, 1, len(mndt.lbl_bin.classes_), (3, 3, 3),
              (64, 64, 128, 256), reg=0.0005)

# Compile the model with specified loss, optmizr, and metrics
model.compile(loss="categorical_crossentropy",
optimizer=optmzr,
        metrics=["accuracy"])

# Define the checkpoint path and directory
chckpnt_pth = "/Users/nikhilkushwaha/ocr/CombineDatasetModels/training_1/cp.h5"
chckpnt_drctry = os.path.dirname(chckpnt_pth)

# Define the callback to save the model's weights
chckpnt_cllbck = tf.keras.callbacks.ModelCheckpoint(filepath=chckpnt_pth, save_weights_only=False,
verbose=1)

# Check if the model wieghts have already been saved or not

if os.path.exists(chckpnt_pth):
    # Load the saved model weights
    model = load_model(chckpnt_pth)
    print("Loaded model weights")

# Train the model and save the
weights hstry = model.fit(
    mndt.augmented.flow(mndt.trn_dta_X, mndt.trn_dta_Y,
    batch_size=btch_sze), validation_data=(mndt.tst_dta_X, mndt.tst_dta_Y),
    steps_per_epoch=len(mndt.trn_dta_X) // btch_sze,
    epochs=nm_epchs,
    class_weight=mndt.class_weight,
    verbose=1,
    callbacks=[chckpnt_cllbck])

# Save the final trained model
save_path = '/Users/nikhilkushwaha/ocr/CombineDatasetModels/final_model_resnet/
Combined_Resnet_20_Epochs.h5'
model.save(save_path)
```

And choosing the **loss function** of the model as **"categorical_crossentropy"**, **"Adam"** as **optimizer** with **learning rate =0.01** and **"accuracy"** as evaluation metrics.

After compiling the model, taking **batch size = 256**. The model is needed to be load the weights for training of the model. Then the model start training using the **model.fit** and then saving the trained model in **.h5** format file [7].

8.2.2.8 *Evaluation of trained model*

Then after training the model three times with epoch = 20, epoch = 10, epoch = 10, the validation accuracy of the model equals 90.21 and a validation loss = 2.45 is achieved. So on evaluating the model using metrics we have the combined f1 score of the model as evaluation metrics.

8.2.2.8.1 Python code for evaluation of model

```
import save
import ocr # Importing the 'ocr' trained model
import mnistdataset as mndt
import numpy as np #Numpy is for handling large multi-dimesional arrays and matrices

from sklearn.metrics import classification_report
 # Importing the 'classification_report' function from the 'sklearn.metrics' module to generate the
classification reports

lblNms = "0123456789ABCDEFGHIJKLMNOPQRSTUVWXYZabcdefghijklmnopqrstuvwxyz"
lbl_Nms = [l for l in lblNms]
print("[INFO] evluatng ntwrk ")
prdctions = save.model.predict(mndt.tst_dta_X, batch_size=ocr.btch_sze)

print(classification_report(mndt.tst_dta_Y.argmax(axis=1),
  prdctions.argmax(axis=1), target_names=lbl_Nms,zero_division=0))
```

8.2.2.8.2 Output for the evaluation of model
Here (figure 8.2), it is shown that the f1-score of each label of the model and the accuracy = 0.89, macro avg = 0.73 and weighted avg = 0.89.

8.3 Model deployment

Now, the model is trained enough to get deployed so that it can be used to convert the text content of the image to text format. First it is required to take the input image, which contains something written in English Language (also numerically from 0–9) [8]. Then it is needed to perform some actions with the image so that model will understand and read the image to predict the required output.

8.3.1 Steps for prediction

```
from imutils.contours import sort_contours
 # 'sort_contours' function from the 'imutils.contours' module for sorting contours
import numpy as np
 # 'numpy' library for numerical operations
import argparse
 # 'argparse' module for parsing command-line arguments
import imutils
 # 'imutils' library used for image processing convenience functions
import cv2
 # cv2' module from OpenCV for computer vision operations
from keras.models import load_model
 # 'load_model' function from the 'keras.models' module for loading pre-trained models
```

1. First, read the input and then use 'cv2.imread()' function.
2. Convert the image to grayscale using 'cv2.cvtColor()' function, as the grayscale images simplify further processing steps.
3. Then apply the Gaussian blur to the grayscale image using 'cv2.GaussianBlur' function. We are applying the Gaussian blur to reduce the noise and smooth out the image. This also helps in enhancing the accuracy of edge detection.
4. Then apply the Canny edge detection to the blurred imaged(edged) as it helps to detect the sharp changes in the intensity, which helps to identify the edges of the characters of the images by using 'cv2.Canny()'.

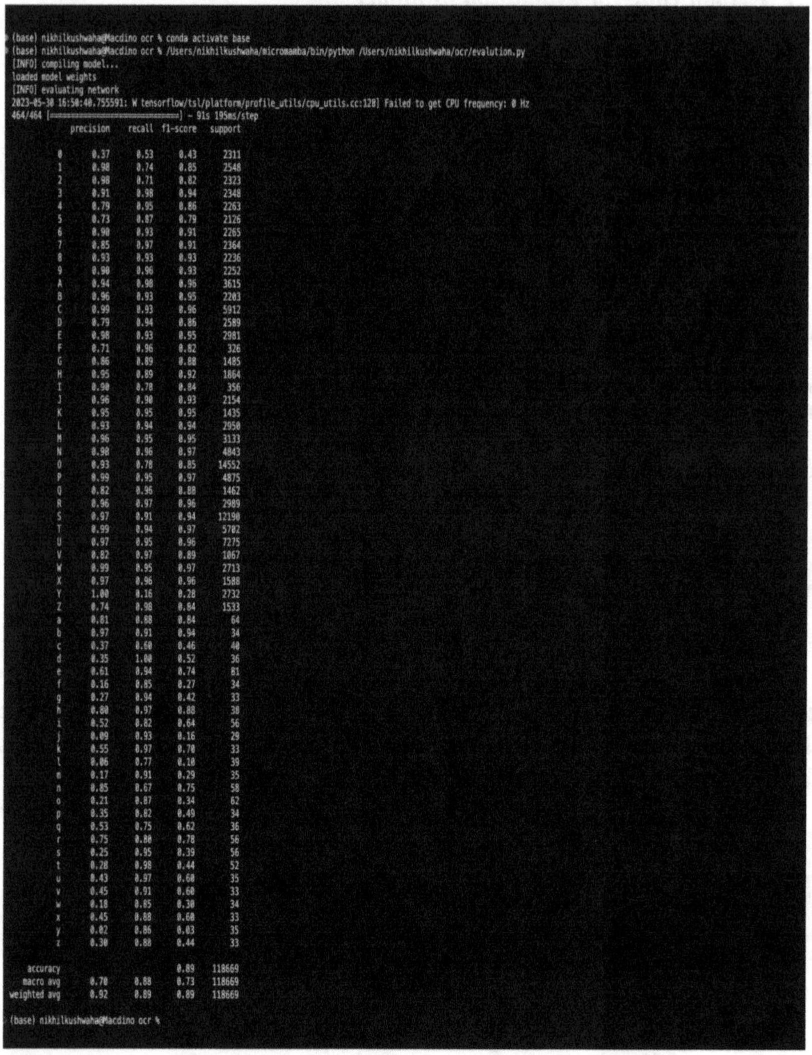

```
(base) nikhilkushwaha@Macdino ocr % conda activate base
(base) nikhilkushwaha@Macdino ocr % /Users/nikhilkushwaha/micromamba/bin/python /Users/nikhilkushwaha/ocr/evaluation.py
[INFO] compiling model...
loaded model weights
[INFO] evaluating network
2023-05-30 16:50:40.755591: W tensorflow/tsl/platform/profile_utils/cpu_utils.cc:128] Failed to get CPU frequency: 0 Hz
464/464 [==============================] - 91s 195ms/step
              precision    recall  f1-score   support

           0       0.37      0.53      0.43      2311
           1       0.98      0.74      0.85      2548
           2       0.98      0.71      0.82      2323
           3       0.91      0.98      0.94      2348
           4       0.79      0.95      0.86      2263
           5       0.73      0.87      0.79      2126
           6       0.90      0.93      0.91      2265
           7       0.85      0.97      0.91      2364
           8       0.93      0.93      0.93      2236
           9       0.90      0.96      0.93      2252
           A       0.94      0.90      0.96      3615
           B       0.96      0.93      0.95      2203
           C       0.99      0.93      0.96      5912
           D       0.79      0.94      0.86      2589
           E       0.98      0.93      0.95      2981
           F       0.71      0.96      0.82       326
           G       0.86      0.89      0.88      1485
           H       0.95      0.89      0.92      1864
           I       0.90      0.78      0.84       356
           J       0.96      0.90      0.93      2154
           K       0.95      0.95      0.95      1435
           L       0.93      0.94      0.94      2950
           M       0.96      0.95      0.95      3133
           N       0.90      0.96      0.97      4843
           O       0.93      0.78      0.85     14552
           P       0.99      0.95      0.97      4875
           Q       0.82      0.96      0.88      1462
           R       0.96      0.97      0.96      2989
           S       0.97      0.91      0.94     12190
           T       0.99      0.94      0.97      5702
           U       0.97      0.95      0.96      7275
           V       0.82      0.97      0.89      1067
           W       0.99      0.95      0.97      2713
           X       0.97      0.96      0.96      1588
           Y       1.00      0.16      0.28      2732
           Z       0.74      0.98      0.84      1533
           a       0.81      0.88      0.84        64
           b       0.97      0.91      0.94        34
           c       0.37      0.60      0.46        40
           d       0.35      1.00      0.52        36
           e       0.61      0.94      0.74        81
           f       0.16      0.85      0.27        34
           g       0.27      0.94      0.42        33
           h       0.80      0.97      0.88        38
           i       0.52      0.82      0.64        56
           j       0.09      0.93      0.16        29
           k       0.55      0.97      0.70        33
           l       0.06      0.77      0.10        39
           m       0.17      0.91      0.29        35
           n       0.85      0.67      0.75        58
           o       0.21      0.87      0.34        62
           p       0.35      0.82      0.49        34
           q       0.53      0.75      0.62        36
           r       0.75      0.80      0.78        56
           s       0.25      0.95      0.39        56
           t       0.28      0.98      0.44        52
           u       0.43      0.97      0.60        35
           v       0.45      0.91      0.60        33
           w       0.18      0.85      0.30        34
           x       0.45      0.88      0.60        33
           y       0.02      0.86      0.03        35
           z       0.30      0.88      0.44        33

    accuracy                           0.89    118669
   macro avg       0.70      0.88      0.73    118669
weighted avg       0.92      0.89      0.89    118669

(base) nikhilkushwaha@Macdino ocr %
```

Figure 8.2. Evaluation metrics of OCR model.

5. Then it is needed to find the contours in the edged image. Contours are the continuous lines that form the boundaries of objects in the images. By using '**cv2.findContours()**' function we can do this.

6. Then it is needed to filter and sort the contours.

7. Then it is required to create an empty list to store the characters and then iterate over the contours.

8. Then we need to compute the bounding box of the contour to extract the region where the character is present. By using '**cv2.boundingRect()**' we can do this.

9. Then the character region is extracted from the grayscale image based on the bounding box coordinates. Then, a binary thresholding technique ('**cv2.threshold()**') is applied to make the character appear white on a black background.

8.3.2 Python code for prediction of image

```python
from imutils.contours import sort_contours
# 'sort_contours' function from the 'imutils.contours' module for sorting contours
import numpy as np
# 'numpy' library for numerical operations
import argparse
# 'argparse' module for parsing command-line arguments
import imutils
# 'imutils' library used for image processing convenience functions
import cv2
# cv2' module from OpenCV for computer vision operations
from keras.models import load_model
# 'load_model' is the func from the 'keras.models' mod for loadng pre-trained models

#Load the pre-trained model
trned_mdel = load_model('/Users/nikhilkushwaha/ocr/CombineDatasetModels/final_model_resnet/
Combined_Resnet_20_Epochs.h5')


def prdct_imge(imge_path):
 imge = cv2.imread(imge_path)
 #change the image to grayscale
 gry_imge = cv2.cvtColor(imge,
 cv2.COLOR_BGR2GRAY) # using Gaussian blur to
 reduce the noise
 blrrd_imge = cv2.GaussianBlur(gry_imge, (5, 5), 0)
 #using canny edge detection
 edgd_imge = cv2.Canny(blrrd_imge, 30, 150)
 #searching contours in edged transformed image
 cnts = cv2.findContours(edgd_imge.copy(), cv2.RETR_EXTERNAL,
  cv2.CHAIN_APPROX_SIMPLE)
 cnts = imutils.grab_contours(cnts)
 # sorting the contours of the edged transfor image left to
 right cnts = sort_contours(cnts, method="left-to-right")[0]
 chrctrs = []


 # process each contours
 for c in cnts:
  (x, y, w, h) = cv2.boundingRect(c)
  if (w >= 5 and w <= 150) and (h >= 15 and h <= 120):
   roi = gry_imge[y:y + h, x:x + w]

   # using thresholding so that to change in binary image
   thrsh = cv2.threshold(roi, 0, 255,
   cv2.THRESH_BINARY_INV | cv2.THRESH_OTSU)[1]
   (thrsh_hieght, thrsh_wdth) = thrsh.shape
   # Resizing the image to a fix size
   if thrsh_wdth > thrsh_hieght:
    thrsh = imutils.resize(thrsh, width=32)
   else:
    thrsh = imutils.resize(thrsh, height=32)
   (thrsh_hieght, thrsh_wdth) = thrsh.shape

   # doing padding to achive fixed size
   x = max(0, 32 - thrsh_wdth)
   y = max(0, 32 - thrsh_hieght)
   dlta_x = int( x / 2.0)
   dlta_y = int( y / 2.0)
   pdded = cv2.copyMakeBorder(thrsh, top=dlta_y, bottom=dlta_x,
    left=dlta_x, right=dlta_x,
    borderType=cv2.BORDER_CONSTANT,
```

```
    value=(0, 0, 0))
    pdded = cv2.resize(pdded, (32, 32))
    pdded = pdded.astype("float32")
    pdded = np.expand_dims(pdded, axis=-1)
    chrctrs.append((pdded, (x, y, w, h)))

# extracting bounding boxes of characters
bxs = [bx[1] for bx in chrctrs]

# converting characters to numpy array
chrctrs = np.array([c[0] for c in chrctrs], dtype="float32")

# Make Predictions using the model
prds = trned_mdel.predict(chrctrs)

# process the prediction
lblNms = "0123456789"
lblNms += "ABCDEFGHIJKLMNOPQRSTUVWXYZ"
lblNms = [l for l in lblNms]
otpt=""
for (prd, (x, y, w, h)) in zip(prds, bxs):
  i = np.argmax(prd)
  prob = prd[i]
  lbl = lblNms[i]
  otpt+=lbl

return otpt
print(prdct_imge("/Users/nikhilkushwaha/ocr/contents/test4.jpeg"))
```

10. Then the threshold image is resized to 32×32 dimension using the '**imutils. resize()**' function.
11. Then it is required to pad the image to maintain the aspect ratio and then append the padded character and its bounding box coordinates to the chars list.
12. Then create the chars list of bounding boxes from the chars list.
13. Then convert the chars list to the Numpy array.
14. Then predict the characters using the model.
15. Then it is needed to define the list of the labels for the characters.
16. Iterate over the predictions and the bounding boxes and then compare the predicted character with the label. Then find the labels with the highest probability match with predicted characters using '**np.argmax()**'
17. The label corresponding to the highest probability is retrieved from the **labelNames** list.
18. Then append the labels to the output string.
19. The output string is our final predicted output.

8.4 Model testing

Now the model is trained and deployed so it needs to be tested by using test images as we develop this model to detect the number plate of vehicles. So we need to test with those images.

8.4.1 TEST1

The test image (figure 8.3) is provided as test1.

It is clearly seen that in the image it is written: **MP506PY2056**.

Then on running **prediction.py** file we got figure 8.4.

It is clearly seen the output of the file **prediction.py** is **VVP505V2055**, which depicts that the model is predicting with accuracy more than **90%** based upon the **f1 score** of each character.

8.4.2 TEST2

It is clearly seen that in the image (figure 8.5) it is written: **DH4859CY**.

Then on running **prediction.py** file we got figure 8.6.

It can be clearly seen the output of the file **prediction.py** is **DHQ5995GYV**, which depicts that the model is predicting with accuracy more than **90%** based upon the **f1 score** of each character.

MP 506 PY 2056

Figure 8.3. Test image 1.

Figure 8.4. Output image of Test1.

Figure 8.5. Test image 2.

8.4.3 TEST3

It can be clearly seen that in the image (figure 8.7) it is written: **MP445NCL5698**. Then on running **prediction.py** file we got figure 8.8.

It can be clearly seen that the output of the file **prediction.py** is **NP445NCL55J98**, which depicts that the model is predicting with accuracy more than **90%** based upon the **f1 score** of each character.

8.4.4 TEST4

It can be clearly seen that in the image (figure 8.9) it is written: **MP809CT5987**. Then on running **prediction.py** file we got figure 8.10.

Figure 8.6. Output image of Test 2.

Figure 8.7. Test image 3.

Figure 8.8. Output image of Test3.

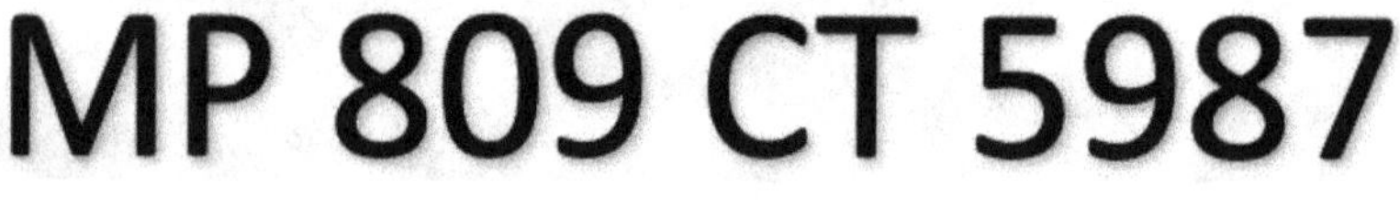

Figure 8.9. Test image 4.

Figure 8.10. Output image of Test4.

It can be clearly seen that the output of the file **prediction.py** is **VVP809CI5981,** which depicts that the model is predicting with accuracy more than **90%** based upon the **f1 score** of each character [3].

8.5 Conclusion

It is well known that OCR can be implemented by using so many methods, techniques and algorithms. In this chapter a convolution neural network is used to build the OCR model with 90% validation accuracy for all the 62 characters including 0–9, a–z, A–Z and that is proven by the above section 8.4 (Model Testing) [1]. In that section it is shown that the model is well trained and able to read the text from the number plate images.

References

[1] Ranjan Jana and Chowdhury R Optical character recognition from text image *Int. J. Comput. Appl. Technol. Res.* **3** 239–43

[2] Sruthi P 2021 A Detailed study and recent research on OCR *Int. J. Comput. Sci. Inform. Sec.* **19** 52–62

[3] Luna A C and Trajano C 2022 License Plate Recognition for Stolen Vehicles Using Optical Character Recognition *ICT Analysis and Applications* (Singapore: Springer Nature)

[4] https://kaggle.com/datasets/sachinpatel21/az-handwritten-alphabets-in-csv-format

[5] https://kaggle.com/datasets/olafkrastovski/handwritten-digits-0-9

[6] https://kaggle.com/datasets/nachiketkanore/english-alphabets-computerized

[7] http://ee.surrey.ac.uk/CVSSP/demos/chars74k-EnglishImg.tgz

[8] Karez A 2016 A detailed analysis of optical character recognition technology *Int. J. Appl. Math. Elec. Comput.* **4** 244

Chapter 9

Automatic COVID-19 identification with a binary neural network using CT images

Rajveer S Lalawat and Varun Bajaj

This chapter classifies the coronavirus disease (COVID-19) with computed tomography (CT) images based on artificial intelligence (AI) and a binary neural network (BNN). The global spread of the COVID-19 pandemic necessitated the development of machine-based disease detection devices in medical imaging. Compared to the traditional x-ray method (which lacks substantial detail), chest CT scan mechanisms produce a more detailed view of the lung, blood vessels and soft tissue. BNNs significantly reduce memory size and usage during forward propagation by replacing mathematical operations with bitwise operations. BNNs have the potential to reduce energy consumption by more than 32 times and the BNN kernel is around 23 times faster than the traditional multiplication kernel. In contrast, the research community is always striving to boost precision and performance parameters of existing diagnostic models and has placed a high reliance on simulations, which were previously used to identify diseases. We aim to develop effective neural networks with binary weights and activations to correctly identify two groups of CT COVID-19 images. The basic idea behind single instruction, multiple data (SWAR) is to concatenate sets of 32 binary parameters with 32-bit registers, which results in a 32-time acceleration on bitwise operations. Field-programmed gate arrays (FPGAs) are highly effective for BNNs due to their reduced memory space.

9.1 Introduction

During the novel COVID-19 pandemic phase numerous studies and actions emerged, offering various solutions to the diverse range of problems that arose as a result of the pandemic. These solutions encompassed multiple fields and addressed various challenges posed by the unprecedented situation [1]. The global spread of the 2019 coronavirus illness (COVID-19) has prompted the need for the creation of techniques based on machines to identify the disease in diagnostic imaging [2].

doi:10.1088/978-0-7503-5924-5ch9

The current COVID-19 epidemic has tested worldwide health-care systems. Furthermore, a lack of instruments and demanding testing requirements make it difficult to screen potentially contagious persons quickly and accurately. As a result, non-laboratory tests such as computer-aided image processing of chest radiography (x-rays) and computed tomography (CT) scans have been used to evaluate lung areas for COVID-19 detection. Scientists and researchers focused on developing vaccines, antiviral medications, and diagnostic tests for COVID-19 [3]. In the context of COVID-19, both CT scans and x-rays have been utilized as imaging modalities to assist in the disease's detection and diagnosis. In order to aid in the analysis and interpretation of these imaging studies, AI-based systems have also been developed [4]. X-rays are more broadly available, more accessible, and less expensive than CT scans, making them the imaging modality of choice in contexts with limited resources, but x-rays provide a two-dimensional view of the chest, limiting the ability to detect subtle lung abnormalities, particularly in the early phases of COVID-19 infection. CT scans have demonstrated greater sensitivity than x-rays for detecting COVID-19 lung abnormalities. This means that CT scans can frequently detect subtle changes in lung tissue caused by infection, even when x-rays appear normal. CT scans give detailed cross-sectional pictures of the lungs, providing a more thorough assessment of the amount and location of lung involvement. CT scans may follow the course of lung anomalies over time, giving important information about disease progression and therapy response. CT scans expose patients to more radiation than x-rays.

Convolutional neural networks (CNNs) have been frequently utilised to analyse and classify COVID-19 CT images. CNNs are deep learning algorithms that are particularly intended to analyse visual data, making them ideal for image identification applications such as medical envision analysis. CNNs can automatically identify and diagnose COVID-19 in CT scans by analysing patterns and characteristics. CNNs for COVID-19 CT images follow these instructions. Training and validation use a large dataset of COVID-19 CT scans, both positive and negative. The dataset should reflect the disease's various phases and presentations.

CT scans may be pre-processed to eliminate noise and artefacts. Resizing, normalisation, and noise reduction are pre-processing methods.

CNN model architecture includes convolutional, pooling, and fully linked layers. Based on the network design (VGG, ResNet, or Inception) or job requirements, the architecture might change.

Labelled CT images train the CNN model. The network learns COVID-19 patterns during training. Back propagation and gradient descent optimise model parameters. A validation dataset is used to test the training model and improve its accuracy and generalisation. Regularisation, dropout, and early ending may avoid over fitting. To evaluate real-world performance, the trained model is tested on unseen CT scans. It may help radiologists find and classify COVID-19 instances in clinical settings or healthcare systems [5–7].

While CNNs have shown to be greatly effective for a variety of image processing applications, they do have certain drawbacks when compared to binary neural networks (BNNs) [8]. CNN models have larger model sizes and memory demands than BNNs. CNNs commonly employ weights and activations with 32-bit or 16-bit precision, which necessitates additional memory for storage and processing. BNNs, on the other hand, employ binary weights and activations, which results in much lower memory needs.

In terms of computing efficiency, BNNs surpass CNNs. When compared to the floating-point operations employed in CNNs, binary operations such as bitwise XOR and bit counting are simpler and quicker to calculate [9]. This reduces processing costs and speeds up inference times in BNNs. When dealing with huge datasets, training CNNs may be computationally and time-consuming [10]. Gradient-based approaches with floating-point weights and activations are required for the optimisation process. Training BNNs, on the other hand, may be less computationally difficult because to the simplicity of binary operations.

9.2 Related work

In the present section, we will discuss the most essential aspects of the top-selected papers utilizing automatic detection of COVID-19 as the artificial intelligence method for diagnosis. Wang et al [11], using COVID-Net, a deep convolution network (CN) expressly designed to find COVID-19 from chest x-rays (CXR) images, detected COVID-19 in this study. The original code for the network that was built has been made public. Ozturk et al [12] used the DarkNet model as a predictor for the YOLO actual time object-detection algorithm to find COVID-19 in CXR visuals. The obtained accuracy was 98.08 % for binary classification and 87.02 % for multiclass. Wang et al [13] thought that it might be possible to get graphical features of COVID-19 from CT scans of the lungs, giving a clinical identification before the viral test that is usually done in a lab. Deep learning strategies were used in order for Xu et al [14] to accomplish their goal of developing an early screening model that could identify COVID-19 from lung CT scans. The dataset used included a total of 618 CT samples that were taken from transverse sections. Song et al [15] used CT scan pictures from 88 subjects with COVID-19, 100 subjects with bacterial pneumonia, and 86 normal individuals to test and train a deep learning-based CT diagnostic system. Ardakani et al [16] estimated the implementation of CNN techniques in routine clinical practise using 10 convolutional neural networks. In their analysis, the authors utilised CT images from 194 subjects (108 COVID-19 and 86 normal). Chen et al [17] developed a CNN model to detect COVID-19 in CT images of high resolution. The authors gathered around 46 000 images from 106 subjects at Wuhan University's Renmin Hospital in China. Deep learning was used to serial CT scan images by Huang et al [18] in order to measure variations in the COVID-19 load. The method included the estimation of quantitative image variables, which generated self-regulation by a deep learning application instrument using chest CT data. Wang et al [19] demonstrated a completely automated deep learning method for COVID-19 diagnosis and risk stratification; 5372 CT scans were utilised in the study. Chinese cities and provinces provided the dataset. The authors used DenseNet121-FPN deep learning to segment CT images for lung masks. Maghdid and his colleagues [20] utilized multiple sources to compile a comprehensive collection of both types of images. The dataset consisted of 170 CXR images and 361 CT scan images, which were then experimented with using a basic CNN and a modified pre-trained AlexNet model. The first model achieved an experimental accuracy of up to 94.1%, while the second model achieved an accuracy of 98%. Amyar *et al* [21] proposed a multitask deep learning (MTL) approach for COVID-19 screening. The MTL architecture proposal was founded on three tasks: classification of COVID-19 versus normal versus other infections, COVID-19 lesion segmentation, and image remodelling. Kassania and her coauthors [22], One of the studies analysed CT scans via the AI Intelligent Assistant Analysis System, which is a programme developed and owned by the organisation. Researchers conducted a study consisting of 20 trials to examine the precision and usefulness of chest CT scans in identifying COVID-19 and their potential applicability in surgical settings. The research evaluated how sensitive CT is for identifying COVID-19-positive persons with and without symptoms [23]. The majority of COVID-19 diagnostic techniques seek for certain CT scan patterns and categorise them as COVID-positive, COVID-negative, COVID, non-COVID, community-acquired pneumonia (CAP), bacterial pneumonia, different types of pneumonia normal, aberrant, infected, not infected [24].

In their study, Xuehai *et al*'s [25] Self-Trans-based model with self-supervised learning to transfer learning was supposed to reduce overfitting. Aayush *et al* [26] set up a way to divide cases into COVID-positive and COVID-negative groups. The suggested model used DenseNet-201 that had already been trained, and shows 97% accuracy. In order to separate the lung areas from CT slices, Wu *et al* [27] used a threshold segmentation technique. They developed a DL model to discriminate between COVID-positive and COVID-negative, and in this test accuracy is 76%.

Wang *et al* [28] developed an inception network used with transfer learning neural networks to classify COVID-19-positive/negative cases. Its accuracy is 76%. Mishra *et al*'s [29] proposed strategy integrates the predictions of each single deep CNN to prevent producing inaccurate predictions about the COVID-19 classes. Shah *et al* [30] developed CT-net10, a deep LM to classify COVID-19-positive and -negative cases. It gets the highest accuracy in VGC-19 and it is performing well in training time. Liu *et al* [31] proposed a lesion-attention deep NN; this technique used multi-label prediction. Ning *et al* [32] developed a VGC-16 based system to distinguish negative/positive cases by employing lung extraction from activation maps, both mild and severe instances. Alshazly *et al* [33] proposed a COVID-Nets based on ResNet and DenseNet to classification of COVID-19, normal viral infections, and healthy classes with results on performance test accuracy of 83%. Zhu *et al* [34] pre-trained a Resnet50 model used in the classification of COVID-19 and normal classes with a result of 93% accuracy.

In this work creating a BNN model using computed tomography (CT) pictures with high accuracy and few configurable parameters is extremely astounding. BNNs are NNs that employ binary weights and activations, which may minimise memory needs and computational complexity when compared to regular neural networks.

9.3 Methodology

The remaining parts of the chapter are organised as follows.

In section 9.2, we discussed data sets and offer a discussion of the COVID-19 diagnosis based on BNN models using CT scan images. In section 9.3, we go through the results of the Python-based Jupyter notebook tools and the developed BNN model. Section 9.4 includes a conclusion.

9.3.1 Datasets

Computed tomography, often known as CT scans, are an effective method for detection of COVID-19 patients while the disease is in its epidemic phase. To address this issue, we developed an open-sourced dataset called COVID-CT. This dataset includes around 350 COVID-19 CT images taken from 215 individuals and also has 463 normal CT scans. A skilled doctor who has been finding and treating COVID-19 cases since the beginning of the epidemic has confirmed that this dataset is useful. A top physician at Tongji Hospital in Wuhan, China, who identified and cared for a large number of COVID-19 patients during the illness's January–April spread, has confirmed that this information is useful and data is available to download from this link: https://github.com/UCSD-AI4H/COVID-CT.

9.3.2 Binary neural networks (BNNs)

CNNs excel in a variety of computer vision applications, including classification and detection of text. A CNN's information can be compressed using dense network architecture. These techniques minimise accuracy loss while minimising the number of parameters and processes by using precision floating point values as weights. Contrarily, information may be reduced by eliminating the activations and weights of a full-precision floating point, which need 4 bytes of storage and are often used. Instead, quantized floating-point integers with lesser precision (one bit of capacity) weights and activations are employed in these systems. A BNN saves up to 32 MB of memory which speeds up the processor by encoding weights and biases by binary values. Additionally, binary operations such as X-NOR and bit count may be used for convolution calculation instead of arithmetic operations. These studies utilised recognised real-valued network topologies to BNNs, such as Alex-Net, Google-Net, or Res-Net, without providing detailed explanations or tests on the design choices [35].

We begin with a quick explanation of the key binary layer development ideas. Then, we go through three frequently used BNN methods: scaling factors, full-precision initial training, and approximation sign function [36]. We did not see the promised result in the accuracy. We show empirically why these strategies are less successful when training from the start. This work encourages us to discover better BNN training methods.

A BNN is a type of NN that takes binary values for both the network weights and the activations. In traditional neural networks, real-valued weights and activations are used, typically represented by floating-point numbers. However, in a BNN, these values are constrained to binary values, usually −1 and 1 or 0 and 1.

The binary representation in BNNs offers several advantages:

1. Memory and computational efficiency: Binary values require less memory to store and lower computational complexity for arithmetic operations compared to real-valued numbers. This makes BNNs particularly useful in contexts with limited resources, like embedded systems or mobile devices.
2. Energy efficiency: BNNs can be more energy-efficient compared to traditional neural networks since binary operations, such as bitwise XOR and bit counting,can be implemented with low-power hardware circuits.
3. Reduced model size: By using weights and activations, the size of the NN model is significantly reduced, and is easier to store and deploy.
4. Increased robustness: Binary values are less sensitive to noise and perturbations, which can make BNNs more robust in certain scenarios, such as in environments with high levels of interference or limited precision hardware.

However, training binary neural networks can be challenging. Traditional optimization techniques, such as gradient descent, cannot be directly applied due to the discontinuity of binary values. Several methods have been developed to tackle those issues, including the use of approximations, regularisation techniques, or special training algorithms like Binary Connect or XNOR-Net [37].

Binary neural networks have found applications in numerous domains, such as voice recognition, computer vision, and language processing. While they may not achieve the same level of accuracy as their real-valued counterparts, BNNs provide a trade-off between performance and efficiency, making them suitable for specific use cases [38].

Convolutional layers are commonly used in deep learning architectures as shown in Fig. 9.1 for processing and analysing data with spatial relationships, such as images. They apply filters (also known as kernels) to input data, performing a series of local computations to extract features (figure 9.1).

9.4 Binary layers implementation

The Sign function for binary activation converts floating-point numbers to binary ones. G (weight/activation) values are binarized using deterministic and stochastic functions, respectively. [39]. The Sign function is defined as follows:

$$G^b = \text{Sign}(G) = \begin{cases} +1 & \text{if } G \geqslant 0, \\ -1 & \text{otherwise} \end{cases} \tag{9.1}$$

The technique makes use of a straight-through estimator (STE), but it also has the additional feature of cancelling the gradients if the inputs are of an excessively high size, as Hubara and his colleague [12] suggested. Let's say that c is the objective function, R_i is an input real integer, and R_o is an output binary number that ranges from −1 to 1.

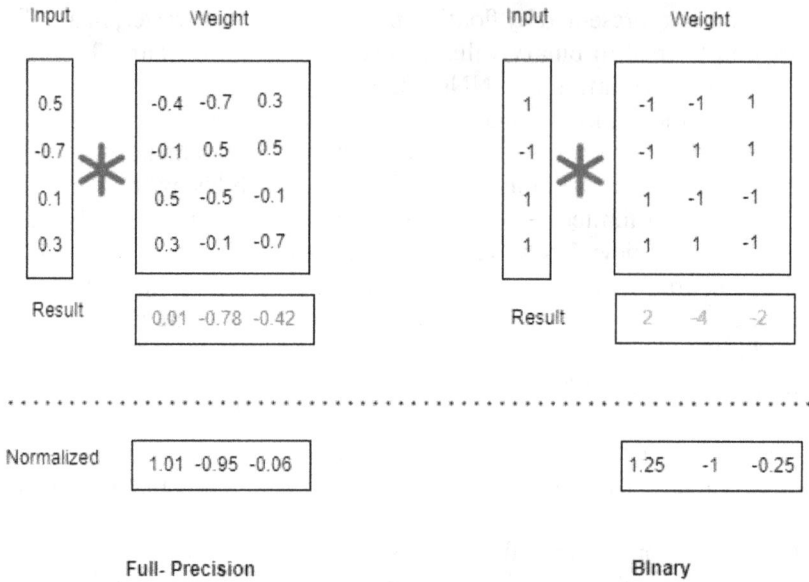

Figure 9.1. Convolution operation method for CNN and BNN model.

$$G^b = \begin{cases} +1 \text{ with probability } p = \sigma(G), \\ -1 \text{ with probability } 1 - p, \end{cases} \qquad (9.2)$$

$$\sigma(G) = \max(0, \min\left(1, \frac{G+1}{2}\right)) \qquad (9.3)$$

In most instances, the deterministic function is used; however, there are a few trials in which the stochastic function is utilised with activations.

In addition to binarizing the weights and activations, there are two more critical aspects that must be present for BNNs to function properly:

Real-valued weights are required in order for optimizers to function properly, and these weights must be acquired using real-valued variables. In spite of the fact that we employ binary weights and activations, the optimisation process makes use of real-valued weights [40].

When attempting to encode binary data using either deterministic or stochastic functions, we encounter an additional challenge. During the process of back propagation, the derivative of that function becomes zero, which means that the gradient as a whole becomes zero as well. Therefore, we are able to make use of the saturated STE, that was first presented by Hinton and investigated by Bengio. In saturated signal transformation encoding (STE), the derivative of the signum function is modified and substituted with a different function. This is a shorthand way of saying that the derivative zero is changed to identity (1) whenever x=1. Therefore, when x is too big, it eliminates the gradient since the derivative is 0 in this case (figure 9.2).

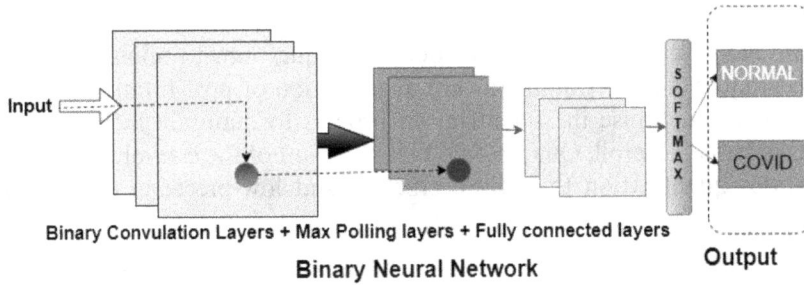

Figure 9.2. Block diagram representation of proposed framework.

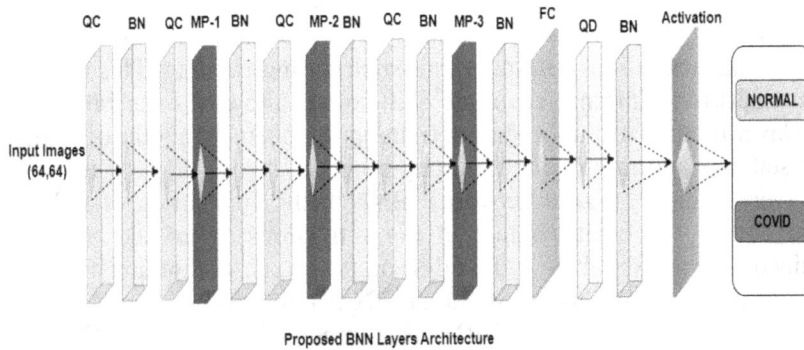

Proposed BNN Layers Architecture

Figure 9.3. Layers representation of proposed framework.

A BNN is a type of NN that uses binary values for both the network weights and the activations. In conventional neural networks, weights and biases are typically represented as real numbers with a precision ranging from 8 to 32 bits, commonly expressed in floating-point notation. However, in a BNN, these values are constrained to binary values, usually -1 and 1 (figure 9.3).

9.4.1 QuantConv2D (QC)

QuantConv2D (QC), as the name suggests, introduces quantization into the convolutional layer. Quantization is the method of reducing the range of values in a neural network, often to lower-bit representations (such as 8-bit or binary values). It is typically done to improve computational efficiency, reduce memory requirements, and enable deployment on resource-constrained devices. QuantConv2D operates similarly to a standard convolutional layer but uses quantized weights and activations. The activations and weights are represented by discrete values instead of continuous values. This minimises the memory impact and operational complexity of the layer.

The specific quantization scheme used in QuantConv2D can vary. It may involve techniques like fixed-point quantization, where real-valued weights and activations are quantized to a fixed number of bits. Alternatively, binary quantization can be employed, where weights and activations are limited to binary values.

The quantization process introduces some loss of information due to the reduced precision. However, modern quantization techniques aim to minimise this loss by using techniques such as training-aware quantization or post-training quantization. These methods optimise the quantization process to maintain model accuracy as much as possible. Overall, QuantConv2D is a variant of the convolutional layer that incorporates quantization to achieve efficient and low-precision computations in convolutional neural networks.

9.4.2 Batch normalization (BN)

BN is a technique commonly used in neural networks, including binary neural networks (BNNs), to normalize the activations of a layer.

In traditional neural networks, BN operates by normalizing the mean and variance of the activations within a mini-batch during training. This normalization helps in addressing internal covariate shift, often known as the phenomenon of altering layer input distribution during training. By normalizing the activations, BN aims to stabilize and improve the training process.

However, in binary neural networks (BNNs), the application of BN is slightly different due to the binary nature of the activations. Since the activations are constrained to binary values (-1 and 1 or 0 and 1), computing the mean and variance in the same way as in traditional BN is not feasible.

Binary batch normalization (BBN) is a modified version of BN specifically designed for BNNs. Instead of computing the mean and variance, BN performs a binarization operation on the inputs. This binarization typically involves thresholding the activations to the binary values of -1 and 1 or 0 and 1.

The BN operation in BNNs can be as follows:
1. Compute the mean and variance of the activations (which are binary values) based on the specific binarization scheme used.
2. Normalize the binary activations by deducting the mean and dividing by the square root of the variance.
3. Apply a scale factor and shift parameter to the normalized activations to provide additional flexibility to the layer.

By applying BN, BNNs can benefit from the regularization and stabilization effects of BN, even with binary activations. BN helps to mitigate the training challenges associated with binary weights and activations and also improve the performance and convergence of binary neural networks.

9.4.3 Max pooling layers (MPLs)

The max pooling layer (MPL) is a widely employed component in CNNs. In CNNs or BNNs max pooling is a technique that downsamples the input by partitioning it into non-overlapping rectangular regions and outputting the maximum value from each region. This helps in reducing the spatial dimensions of the input while preserving the most prominent features.

MPLs have a fixed pool size (e.g., 2×2 or 3×3) and a stride that determines the step size used to move the pooling window across the input. The most common choice is a pool size of 2×2 with a stride of 2, which reduces the spatial dimensions of the input by half.

The main objective of max pooling is to reduce the dimensions of the input images while retaining the most important features. By downsampling the input, max pooling helps in reducing the computational complexity and parameter count of the subsequent layers. Additionally, max pooling can enhance the network's robustness to small variations or translations in the input data. Moreover, max pooling provides some degree of translation invariance. It means that even if the input features slightly shift, the pooling operation can still capture the strongest features within the local region. This property can be advantageous in tasks where the exact spatial location of features is not crucial, such as image classification.

MPL are typically applied between convolutional layers in CNN architectures. They are usually applied after one or several CL to reduce the number of map elements featured progressively. The parameters of an MPL include the window size and stride. The window size determines the size of the local region from which the maximum value is selected. Common choices are 2×2 or 3×3 windows. The stride finds the step size at which the window moves across the input. A stride of two reduces the output size by half, while a stride of 1 retains more spatial information. It's worth noting that max pooling is a non-parametric operation and does not involve any learnable parameters. The network learns the pooling behaviour and the pooling regions based on the training data during the overall training process.

9.4.4 Fully connected layers (FCLs)

FCLs are a type of layer commonly used in neural networks. Each neuron is linked to every neuron in the preceding layer in a completely connected layer, resulting in a dense or fully connected structure. These layers play a crucial role in learning complex relationships and to predict the features based on the learning.

The FCLs take the flattened feature vector as input and learn complex combinations of features to make predictions. The completely connected layer gets information from all neurons in the previous layer and applies an activation function (e.g., ReLU or sigmoid) to introduce non-linearity. Total neurons in each FCL and the depth of the network depend on the specific problem and architecture [41]. The final fully connected layer typically has the same number of neurons as the desired output, and the activation function applied depends on the nature of the problem (e.g., softmax for classification tasks).

The weights of the FCL are adjusted during the training process using optimization algorithms (e.g., back propagation) to minimize a defined loss function. This allows the network to learn meaningful representations and make accurate predictions based on the learned features.

9.4.5 Quant dense layer (QD)

In a quant dense layer, both the weights and activations are quantized to a smaller number of bits (e.g., 4-bit or 8-bit) instead of the typical 32 floating number representation. This reduces the memory footprint and computation cost of the layer, while still allowing the network to achieve reasonable accuracy. Computational effectiveness and spatial invariance are key aspects of convolutional neural networks.

The quantized dense layers take the flattened feature vector as input and learn complex combinations of features using reduced bit representation. Each neuron in the quantized dense layer receives input from every neuron in the previous layer and applies non-linearity function for activation.

The number of neurons in each quantized dense layer, the quantization bit size, and the depth of the network depend on the specific problem and architecture. The final quantized dense layer typically has the same number of neurons as the desired output, and the activation function applied depends on the problem that occurs. The weights of the quantized dense layers are adjusted during the training process using optimization algorithms (e.g., back propagation) to minimize a defined loss function. This allows the network to learn meaningful representations and make accurate predictions based on the reduced-bit learned features.

9.4.6 Activation function

An activation function applied to mathematics form the outcome to a neuron or a set of neurons in a NN. It introduces non-linearity to the network, allowing learning complex patterns and making nonlinear transformations to the input data. Activation functions play a important role in determining the outcome of a neuron and, ultimately, the overall performance of a neural network. Here are some commonly used activation functions.

Sigmoid

The sigmoid function, or logistic function, connects to the input range from 0 and 1. It has an S shape function curve and is given by the formula:

$$f(R) = 1/(1 + \exp(-R))$$

The sigmoid function was generally used in the past but has fallen out of favour for hidden layers due to the problem of vanishing gradient. Furthermore, it is used in the last layer of a neural network for tasks involving binary decision-making, such as choosing between two possibilities (e.g., yes or no), where it is necessary to calculate the probabilities associated with each option.

Softmax

The softmax function is also applied in the last layer of an NN to different classification problems. The process takes vector numbers in real form as input and normalizes them into a probability distribution. It ensures that the output values sum up to 1, making it suitable for representing class probabilities. The softmax function is given by:

$$f(R) = \exp(R)/\text{sum}(\exp(R))$$

It is often used in conjunction with the cross-entropy loss function.

9.5 Result

BNNs present unique training challenges due to the discrete and non-differentiable nature of binary operations. Specialized techniques, such as approximations, relaxation methods, or gradient estimations, need to be employed to train BNNs effectively. These challenges require careful consideration to achieve optimal performance (figure 9.4 and figure 9.5 and tables 9.1–9.3).

9.5.1 Python pseudo code

- The kernel in a binary neural network refers to the convolutional kernel or filter used in the convolutional layers of the network. In convolutional neural networks (CNNs), convolutional kernels are small matrices applied to the input data to extract features and perform spatial filtering.
- In binary neural networks, the convolutional kernels are also binary. Instead of using real-valued weights, the convolutional kernels consist of binary values that are either −1 or +1. These binary convolutional kernels are used to perform binary convolutions with the binary input data. The binary convolutions involve simple operations such as bitwise XNOR (exclusive NOR) and pop count (counting the number of set bits).
- Binary neural networks with binary kernels have gained attention due to their potential for efficient computation and low memory requirements. They can

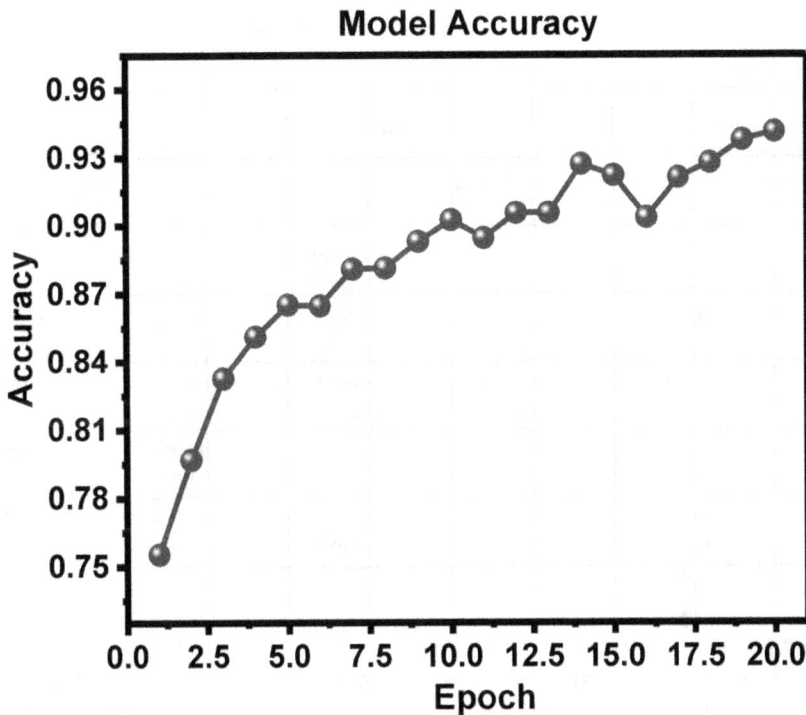

Figure 9.4. Model accuracy with 20 epochs.

Model Loss

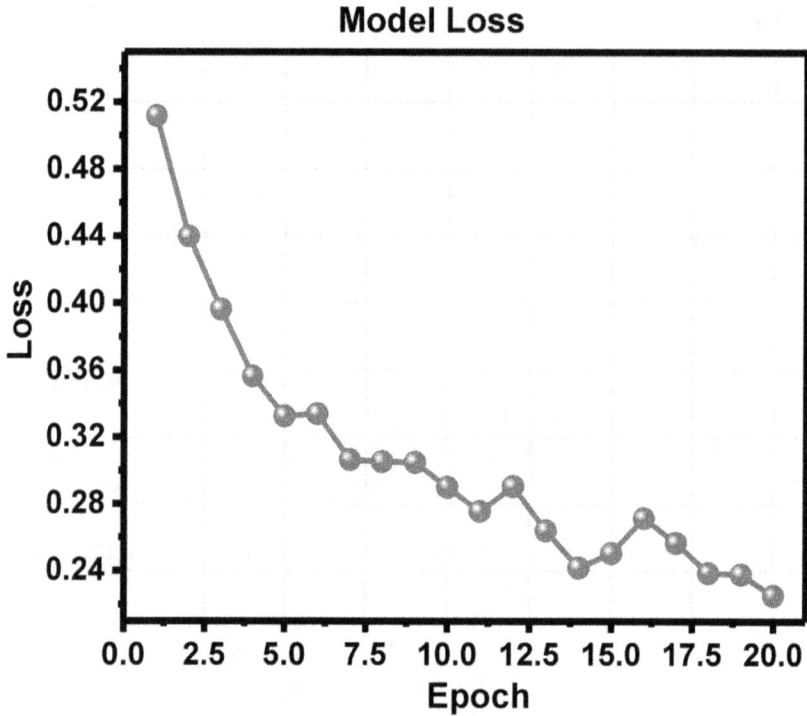

Figure 9.5. Model loss with 20 epochs.

Table 9.1. Top five validation accuracy for 10 epochs (fold 1–5).

Epochs No.	Accuracy-1	Accuracy-2	Accuracy-3	Accuracy-4	Accuracy-5
1	0.9549	0.9790	0.9800	0.9935	0.9895
2	0.9680	0.9765	0.9880	0.9875	0.9965
3	0.9735	0.9810	0.9820	0.9895	0.9955
	0.9725	0.9870	0.9865	0.9955	0.9915
4	0.9680	0.9880	0.9930	0.9945	0.9945
5	0.9680	0.9840	0.9915	0.9920	0.9980
6	0.9690	0.9800	0.9910	0.9925	0.9985
7	0.9730	0.9850	0.9905	0.9960	0.9945
8	0.9805	0.9880	0.9890	0.9955	0.9945
9	0.9700	0.9890	0.9945	0.9955	0.9935
10	0.9775	0.9815	0.9890	0.9950	0.9980

be particularly useful in scenarios with limited computational resources, such as mobile and embedded devices. By using binary values for both weights and activations, binary neural networks can achieve faster inference and require less memory storage compared to traditional neural networks.

Table 9.2. Total parameters used in proposed model.

S. No.	Total parameter	Trainable parameter	Non-trainable parameter
1	14 475 398	14 471 298	4100

Table 9.3. Comparison of parameters with different models (in millions).

S No.	Alex-Net	VGG16	Res-Net	MobileNet	Proposed
1	62.3	138	11	13	1.14

Ten-fold validation Python pseudo code:

load the required libraries (some library mention below)

Numpy

Tensorflow

Sklearn matrics used confusion matrix accuracy, f1_score, precision, and

Sklearn with model selection StratifiedKFold

For image PIL import

Import random , import sys, import cv2, import os

Import larq as lq (for BNN)

models Sequential form Tensorflow

Image Data Generator from keras pre-processing

Import ConfigProto From tensorflow compat.v1

Callbacks TensorBoard, ModelCheckpoint From tensorflow keras

load the image into system memory

```
from zipfile import ZipFile

data="Ct_covid.zip"

with ZipFile (data,'r') as zip:

 zip.extractall()

 print ('Done')
```

Code logic will include the epochs and the validation accuracy

```
FOLD = StratifiedKFold (splits=10, shuffle=True)

FOLD.get_splits (P, Q)

Fold number = 0 (initially)

For Train index, Validation Index in FOLD.split (P, Q):

   Transfer folders ('validation', 'Train', 1.0)

   Fold number += 1

   Print ("Results", number of folds)

   P, train, P, validation = P [Train index], P [validation]

   Q, train, Q, validation = Q [Train index], Q [validation]

   Validation data generator = Image Data Generator (re-scale=1/255)

   Test data generator = Image Data Generator (re-scale=1/255)
```

Build and train the model

```
Batchs size =32
Epoch = 20
Num of channels = 1
Number of class labels = len(class_labels)

Input image row, input image column = 64, 64 (image size of Input )


Train path = os path join(dataset folder, 'Training')
  Validation  path  =  os  path  join  (dataset  folder,
'Validation')
```

```
Test path = os path join (dataset folder, 'Testing')
def get_model():
    Model = tf keras models Sequential ([
        Iq layers Quant Conv2D (128, 3,
        Quantizer kernel ="ste sign",
        Constraint kernel ="weight clip",
        Use bias=False,
        Input size=(64, 64, 1)),
    tf.keras.layers.BatchNormalization(moment=0.999,
scale=False),
    lq.layers.QuantConv2D(128,    3,    padding="same",
**kwargs),
    tf.keras.layers.MaxPool2D(pool_size=(2,        2),
strides=(2, 2)),
    tf.keras.layers.BatchNormalization(moment=0.999,
scale=False),
    lq.layers.QuantConv2D(256,    3,    padding="same",
**kwargs),
    tf.keras.layers.BatchNormalization(moment=0.999,
scale=False),
    lq.layers.QuantConv2D(256,    3,    padding="same",
**kwargs),
    tf.keras.layers.MaxPool2D(pool_size=(2,        2),
strides=(2, 2)),
    tf.keras.layers.BatchNormalization(momentum=0.999,
scale=False),
    lq  layers  QuantConv2D(512,  3,  padding="same",
**kwargs),
    tf keras layers Batch Normalization(momentum=0.998,
scale=False),
    tf   keras   layers   MaxPool2D(pool_size=(2,  2),
strides=(2, 2))
'
    tf keras layers Batch Normalization(momentum=0.998,
scale=False),
    tf keras layers Flatten(),
    lq layers QuantDense(512, **kwargs),
    tf keras layers Batch Normalization(momentum=0.998,
scale=False),
    lq layers QuantDense(2, **kwargs),
    tf keras layers Batch Normalization(momentum=0.998,
scale=False),
    tf keras layers Activation("softmax")
```

```
    ])
       Return model
    kwargs =      dict (quantizer input ="ste sign",
             Quantizer kernel ="ste sign",
             Constraint kernel ="weight clip",
             use bias=False)
    Model = get the model ()
    Model  compile  (optimizer='Adam'  metrics=['accuracy'])
    loss='categorical cross entropy')
    Model summary ()
```

9.6 Conclusion

Binary neural networks have shown potential in terms of decreasing memory requirements and computational complexity while analysing COVID-19 CT images. When compared to standard neural networks, binary neural networks employ binary values for both weights and activations, which may result in quicker inference and less memory storage. It is crucial to highlight, however, that the application of binary neural networks for COVID-19 CT image interpretation is still a growing area with several obstacles to solve. Due to the discontinuous and non-differentiable character of binary operations, training binary neural networks using binary kernels is one of the biggest difficulties. To efficiently train binary neural networks, specialised approaches such as approximations, relaxation methods, or gradient estimates must be used.

Despite these obstacles, preliminary research and tests have shown encouraging outcomes. Binary neural networks have been shown to identify and categorise COVID-19-related anomalies in CT scans. They are a viable option for real-time and resource-constrained applications, such as point-of-care diagnostics or deployment on mobile and embedded devices, because to their computational efficiency and lower memory needs. To optimise binary neural networks for COVID-19 CT image analysis, further research and development are required. This involves investigating better training procedures, assessing performance on bigger datasets, and comparing the findings to classic neural networks and other deep learning methodologies. The use of binary neural networks in medicine, especially COVID-19 diagnosis, is still an active area of study, and future advances may lead to more robust and accurate solutions.

References

[1] Huang C, Wang Y, Li X and Ren L 2020 Clinical features of patients infected with 2019 novel coronavirus in Wuhan, China *Lancet* **395** 497–506

[2] Guan W, Ni Z, Hu Y and Liang W 2019 Clinical characteristics of coronavirus disease in China *N. Engl. J. Med.* **382** 1708–20

[3] Lu H, Stratton C W and Tang Y W 2020 Outbreak of pneumonia of unknown etiology in Wuhan, China: the mystery and the miracle *J. Med. Virol.* **92** 401

[4] Kubina R and Dziedzic A 2020 Molecular and serological tests for COVID-19. A comparative review of sars-cov-2 coronavirus laboratory and point-of-care diagnostics *Diagnosis.* **10** 434

[5] Perumal V, Narayanan V and Rajasekar S J S 2021 Prediction of COVID-19 with computed tomography images using hybrid learning techniques *Dis. Markers* **2021** 5522729

[6] Polsinelli M, Cinque L and Placidi G 2020 A light CNN for detecting COVID-19 from CT scans of the chest *Pattern Recogn. Lett.* **140** 95–100

[7] Iandola F N, Han S, Moskewicz M W and Ashraf K 2016 SqueezeNet: AlexNet-level accuracy with 50x fewer parameters and <0.5 MB model size *arXiv preprint* arXiv:1602.07360

[8] Bethge J, Yang H, Bornstein M and Meinel C 2019 Back to simplicity: How to train accurate BNNs from scratch? *arXiv preprint* arXiv:1906.08637

[9] Rastegari M, Ordonez V, Redmon J and Farhadi A 2016 Xnornet: imagenet classification using binary convolutional neural networks *European Conference on Computer Vision* (Berlin: Springer) 525–42

[10] Wang J, Bao Y, Wen Y and Lu H 2020 Prior-attention residual learning for more discriminative COVID-19 screening in CT images *IEEE Trans. Med. Imag.* **39** 2572–83

[11] Wang L, Lin Z Q and Wong A 2020 COVID-net: a tailored deep convolutional neural network design for detection of COVID-19 cases from chest x-ray images *Sci. Rep.* **10** 19549

[12] Ozturk T, Talo M, Yildirim E A, Baloglu U B, Yildirim O and Acharya U R 2020 Automated detection of COVID-19 cases using deep neural networks with x-ray images *Comput. Biol. Med.* **121** 103792

[13] Wang G, Liu X, Shen J, Wang C, Li Z, Ye L, Wu X, Chen T, Wang K, Zhang X *et al* 2021 A deep-learning pipeline for the diagnosis and discrimination of viral, non-viral and COVID-19 pneumonia from chest x-ray images *Nat. Biomed. Eng.* **5** 509–21

[14] Xu X, Jiang X, Ma C, Du P, Li X, Lv S, Yu L, Ni Q, Chen Y, Su J *et al* 2020 A deep learning system to screen novel coronavirus disease 2019 pneumonia *Engineering* **6** 1122–9

[15] Song Y, Zheng S, Li L, Zhang X, Zhang X, Huang Z, Chen J, Wang R, Zhao H, Chong Y *et al* 2021 Deep learning enables accurate diagnosis of novel coronavirus (COVID-19) with CT images *IEEE/ACM Trans. Comput. Biol. Bioinform.* **18** 2775–80

[16] Ardakani A A, Kanafi A R, Acharya U R, Khadem N and Mohammadi A 2020 Application of deep learning technique to manage COVID-19 in routine clinical practice using CT images: results of 10 convolutional neural networks *Comput. Biol. Med.* **121** 103795

[17] Chen J, Wu L, Zhang J, Zhang L, Gong D, Zhao Y, Chen Q, Huang S, Yang M, Yang X *et al* 2020 Deep learning-based model for detecting 2019 novel coronavirus pneumonia on high-resolution computed tomography *Sci. Rep.* **10** 19196

[18] Huang L, Han R, Ai T, Yu P, Kang H, Tao Q and Xia L 2020 Serial quantitative chest CT assessment of COVID-19: a deep learning approach *Radiol. Cardiothorac. Imaging* **2** e200075

[19] Wang S, Zha Y, Li W, Wu Q, Li X, Niu M, Wang M, Qiu X, Li H, Yu H *et al* 2020 A fully automatic deep learning system for COVID-19 diagnostic and prognostic analysis *Eur. Respir. J.* **56** 2000775

[20] Maghdid H S, Asaad A T, Ghafoor K Z, Sadiq A S, Mirjalili S and Khan M K 2021 Diagnosing COVID-19 pneumonia from x-ray and CT images using deep learning and transfer learning algorithms *Proc. Multimodal Image Exploitation and Learning, Online, 12–16 April 2021* **vol. 11734** (Bellingham, WA: SPIE) pp 99–110

[21] Amyar A, Modzelewski R, Li H and Ruan S 2020 Multi-task deep learning based CT imaging analysis for COVID-19 pneumonia: classification and segmentation *Comput. Biol. Med.* **126** 104037

[22] Kassania S H, Kassanib P H, Wesolowskic M J, Schneidera K A and Detersa R 2021 Automatic detection of coronavirus disease (COVID-19) in x-ray and CT images: a machine learning based approach *Biocybern. Biomed. Eng.* **41** 867–79

[23] Zhang H T, Zhang J S, Zhang H H, Nan Y D, Zhao Y, Fu E Q, Xie Y H, Liu W, Li W P, Zhang H J *et al* 2020 Automated detection and quantification of COVID-19 pneumonia: CT imaging analysis by a deep learning-based software *Eur. J. Nucl. Med. Mol. Imaging* **47** 2525–32

[24] Shao J M, Ayuso S A, Deerenberg E B and Elhage S A 2020 A systematic review of CT chest in COVID-19 diagnosis and its potential application in a surgical setting *Colorectal Dis* **22** 993–1001

[25] Yang X, He X, Zhao J and Zhang Y 2020 COVID-CT-dataset: a CT scan dataset about COVID-19 arXiv preprint arXiv:2003.13865

[26] Jaiswal A, Gianchandani N, Singh D and Kumar V 2020 Classification of the COVID-19 infected patients using DenseNet201 based deep transfer learning *J. Biomol. Struct. Dyn.* **39** 1–8

[27] Wu X, Hui H, Niu M and Li L 2020 Deep learning-based multi-view fusion model for screening 2019 novel coronavirus pneumonia: a multicentre study *Eur. J. Radiol.* **128** 109041

[28] Wang S, Kang B, Ma J and Zeng X 2021 A deep learning algorithm using CT images to screen for corona virus disease (COVID-19) *Eur. Radiol.* **31** 6096–104

[29] Mishra A K, Das S K, Roy P and Bandyopadhyay S 2020 Identifying COVID-19 from chest CT images: a deep convolutional neural networks based approach *J. Heal. Eng.* **2020** 8843664

[30] Shah V, Keniya R, Shridharani A and Punjabi M 2021 Diagnosis of COVID-19 using CT scan images and deep learning techniques *Emerg. Radiol.* **28** 497–505

[31] Liu B, Gao X, He M and Lv F 2020 Online COVID-19 diagnosis with chest CT images: lesion-attention deep neural networks *MedRxiv*

[32] Ning W, Lei S, Yang J and Cao Y 2020 Open resource of clinical data from patients with pneumonia for the prediction of COVID-19 outcomes via deep learning *Nat. Biomed. Eng.* **4** 1197–207

[33] Alshazly H, Linse C, Abdalla M and Barth E 2021 COVID-nets: deep CNN architectures for detecting COVID-19 using chest CT scans *MedRxiv*

[34] Zhu Z, Zhang X, Tao G and Dan T 2021 Classification of COVID-19 by compressed chest CT image through deep learning on a large patients cohort *Comp. Life. Sci.* **13** 73–82

[35] Zhang J and Verma N 2019 An in-memory-computing DNN achieving 700 TOPS/W and 6 TOPS/mm 2 in 130-nm CMOS *IEEE J. Emerging and Selected Topics in Circuits and Systems* **9** 358–66

[36] Redmon J, Divvala S K, Girshick R B and Farhadi A 2016 You only look once: unified, real-time object detection *IEEE Conf. on Computer Vision and Pattern Recognition* vol 8 (Piscataway, NJ: IEEE) 779–88

[37] Ren S, He K, Girshick R and Sun J 2015 Faster R-CNN: towards real-time object detection with region proposal networks *Adv. Neural Inf. Process.ing Syst.* **28** 91–9

[38] Szegedy C, Liu W, Jia Y, Sermanet P, Reed S, Anguelov D, Erhan D, Vanhoucke V, Rabinovich A *et al* 2015 Going deeper with convolutions *IEEE Conference on Computer Vision and Pattern Recognition (CVPR)* (Piscataway, NJ: IEEE)

[39] Tang W, Hua G and Wang L 2017 How to train a compact binary neural network with high accuracy *Thirty-First AAAI Conf. on Artificial Intelligence* (Washington, DC: AAAI)

[40] Wan D, Shen F, Liu L, Zhu F, Qin J, Shao L and Tao Shen H 2018 TBN: convolutional neural network with ternary inputs and binary weights *Computer Vision – ECCV 2018* (Berlin: Springer) 322–39

[41] Yang H, Fritzsche M, Bartz C and Meinel C 2017 Bmxnet: an open-source binary neural network implementation based on mxnet *Proc. ACM on Multimedia Conf.* (New York: ACM) pp 1209–12 8

IOP Publishing

Image Processing with Python
A practical approach

Irshad Ahmad Ansari and Varun Bajaj

Chapter 10

A review and implementation of image despeckling methods

Rishab Sarkar and P V Sudeep

Speckle noise is the presence of a granular pattern on images, resulting in low contrast and low resolution images. In the field of medical imaging, crucial diagnostic procedures such as medical ultrasound (US) and optical coherence tomography (OCT) images often encounter significant quality degradation due to the presence of speckle noise and it negatively impacts the accuracy of diagnoses. Consequently, numerous approaches have been proposed in the literature to mitigate speckle noise in medical images. The traditional despeckling filters, employed in the spatial domain and transform domain, require manual parameter settings. On the other hand, data-driven despeckling methods improve the overall quality of medical images through automatic representative feature learning and offer an improved despeckling performance and preservation of vital image details. Recently, deep learning methods have been popular in data-driven approaches and have emerged as a promising solution to deliver superior despeckling results. In this chapter, we present a review of significant image despecking methods developed in the past. Additionally, we delve into the design cycle of deep learning methods for image despecking, with illustrative examples. In this chapter, we used python scripts to implement the despeckling filters. Furthermore, it is important to note that we have included a few sample experiments, enabling a qualitative and quantitative comparison of the results solely for explanatory purposes.

10.1 Introduction

Medical imaging techniques are non-invasive procedures that enable physicians to look inside the body without the need for surgical intervention. Medical imaging has revolutionized the diagnosis and treatment of various medical conditions. X-ray radiography is a popular method that utilizes ionizing radiation for bone imaging. CT scanners utilize x-rays and computer processing to produce detailed cross-sectional

doi:10.1088/978-0-7503-5924-5ch10

images of the internal body. It is commonly used for diagnosing and monitoring conditions such as cancer, heart disease, and traumatic injuries [24]. MRI is an imaging technique that uses a magnetic field and radio waves to create detailed images of the internal body. It is useful in various medical fields, including neurology, orthopedics, cardiology, oncology, and abdominal imaging [26]. PET [10] and SPECT [40] are nuclear medicine imaging techniques that offer unique functional information that complements the anatomical details obtained from other imaging modalities. PET and SPECT scans use a radiotracer that emits positrons and gamma rays, respectively. Both scans involve the injection of a radiotracer into the patient's bloodstream, which accumulates in areas of interest based on metabolic activity. The emitted radiation is detected by specialized cameras, and computer algorithms create detailed images that show the distribution of the radiotracer. PET and SPECT scans are widely used in diagnosing and monitoring various medical conditions, including cancer, cardiovascular diseases, neurological disorders, and psychiatric disorders. Medical optical imaging refers to a group of imaging techniques that use light to visualize and capture images of biological tissues and structures within the body.

OCT employs low-coherence light to effectively capture high-resolution, cross-sectional images of tissues. It is useful in ophthalmology for imaging the retina and identifying conditions like macular degeneration and glaucoma. OCT is also employed in dermatology, cardiology, and gastroenterology, to visualize tissue layers and structures [11]. Compared to OCT, US imaging uses high-frequency sound waves instead of light. Ultrasound is non-invasive, portable, and safe, making it a versatile imaging modality. It is commonly used in obstetrics for prenatal imaging, as well as in cardiology, radiology, and many other medical specialties [22].

However, various factors such as equipment malfunction, electrical interference, patient movement, and variations in tissue properties cause noise during image acquisition [27]. The presence of noise in the acquired images can make it difficult to interpret the images accurately and can potentially lead to incorrect diagnosis or treatment. Noise in medical images can take different forms, such as random variations in pixel intensity, streaks or lines in the image, or distortion and blurring due to motion. Gaussian noise found in x-ray, CT scan, US, MRI is caused by fluctuations in the electrical signals generated by the imaging equipment [16] and Poisson noise found in PET and SPECT occurs due to the statistical nature of photon detection in low-light imaging scenarios [15]. Speckle noise found in US and OCT appears as a granular pattern in images caused by the coherent nature of the imaging system [3]. Managing these various types of noise is essential for accurate image analysis, diagnosis, and treatment planning in medical imaging, necessitating the development of robust noise reduction techniques.

10.2 Speckle noise

Speckle noise is a common phenomenon encountered in images acquired through coherent imaging systems like synthetic aperture radar (SAR), US, OCT, and laser imaging [25]. This granular disturbance originates from the interference of electromagnetic or acoustic waves reflected from the imaged object. When waves interact with a rough surface, they undergo constructive and destructive interference, resulting in the appearance of randomly scattered small bright or dark dots throughout the image [6].

The key characteristic of speckle noise is its non-Gaussian distribution, necessitating the use of more complex statistical models to comprehensively analyze and describe it. The texture-like appearance of speckle noise poses challenges in distinguishing it from other types of image textures or features, further complicating the accurate identification and characterization of tissue structures. Additionally, the appearance and severity of speckle noise can be influenced by various imaging parameters, such as the wavelength of the imaging signal, the size of the imaging aperture, and the number of samples acquired during the imaging process [25]. As a result, developing general-purpose despeckling algorithms that effectively address speckle noise across different imaging modalities remains a challenging task.

10.2.1 Speckle noise characteristics in medical images

Speckle noise exhibits several challenging characteristics, including randomness, correlation, and a texture-like appearance. Removing or filtering speckle noise without compromising the underlying image information is difficult due to its random and unpredictable nature. However, there is some correlation among neighboring pixels, as they tend to exhibit similar levels of noise. This correlation can be utilized to some extent for noise reduction in images [31]. Also, a better statistical characterization of the observed (noisy) images can greatly assist in the development of effective despeckling methods.

In US imaging, speckle noise appears due to the interference of high frequency sound waves that are scattered and reflected by the tissue structures in the body. The echoes generated by these waves interfere with each other and create constructive and destructive interference patterns. Speckle can be categorized into four types: fully-developed, fully resolved, partially-developed, and partially-resolved, based on the scatter number density per cell (SND) [29]. Among these types, the most commonly encountered in practice are fully-developed and fully-resolved speckle. In the case of a high number of scatters, the fully-developed model assumes a Rayleigh distribution for the envelope, which is often employed for characterizing the magnitude of speckle noise using a scale parameter [14]. While the Rayleigh distribution is widely used, the Gamma distribution offers more flexibility by adjusting shape and scale parameters, allowing it to capture the statistical properties of speckle noise beyond what the Rayleigh distribution can handle [34].

In the literature, Nakagami, log-normal, Rician inverse Gaussian (RiIG) and Weibull are a few examples of other probabilistic models that have been attempted to model empirical distributions obtained from the US images. The Nakagami distribution provides a versatile framework for modeling speckle noise, incorporating parameters to control the shape and spread of the noise [19]. The log-normal distribution is often employed to approximate the distribution of speckle noise. It assumes a normal distribution for the logarithm of noise values and enables greater flexibility in modeling skewed or asymmetric noise distributions [18]. Furthermore, the Weibull distribution is occasionally used to represent speckle noise, as it allows adjustment of shape and scale parameters to capture different noise characteristics [13]. So, the choice of a specific distribution depends on the unique characteristics of the speckle noise and the requirements of the particular application or imaging system at hand.

Besides, in OCT scans the speckle noise arises due to the interference of light waves that are scattered by the tissue being imaged [35]. Also, more attempts to model speckle patterns in US images are present in the literature compared to the OCT.

10.2.2 Speckle noise model

The multiplicative model is a widely used mathematical framework to describe speckle noise. According to this model, the observed image $J(x, y)$ at a specific pixel location (x, y) is expressed as the product of the true underlying scene $I(x, y)$ and a multiplicative noise term $N(x, y)$.

$$J(x, y) = I(x, y) \cdot N(x, y) \tag{10.1}$$

Here, $I(x, y)$ represent the actual image at the pixel location (x, y), and $N(x, y)$ represents the speckle noise at that location.

```python
def add_speckle_noise(image, gamma, sigma):
    # Ensure the input image is a numpy array
    image = np.array(image, dtype=float)
    # Generate zero-mean Gaussian noise
    noise = np.random.normal(0, sigma, size=image.shape)
    # Apply speckle noise according to the given mathematical model
    noisy_image = image + (image ** gamma) * noise
    # Ensure pixel values are in the valid range [0, 255]
    noisy_image = np.clip(noisy_image, 0, 255)
    return noisy_image.astype(np.uint8)


# Add speckle noise with sigma = 30
noisy_image1 = add_speckle_noise(img1, gamma=0.5, sigma=30)
noisy_image2 = add_speckle_noise(img2, gamma=0.5, sigma=30)
noisy_image3 = add_speckle_noise(img3, gamma=0.5, sigma=30)
noisy_image4 = add_speckle_noise(img4, gamma=0.5, sigma=30)
```

Recent studies show the Gamma distribution [38] and the Fisher–Tippett distribution [33] are suitable fits to the observed data having speckle noise distribution. So, in [9], the authors provide an alternative speckle noise model as given below:

$$J(x, y) = I(x, y) + I^{\gamma}(x, y) \cdot \eta(x, y) \tag{10.2}$$

In this equation, $\eta(x, y)$ $N{\sim}(0, \sigma^2)$ denotes zero-mean Gaussian noise and the default value for γ is 0.5. This model offers improved capabilities for capturing image statistics by accounting for ultrasound devices and associated image formation processes [9]. Here, the noise component is dependent on the specific characteristics of the image. The python code for adding speckle noise to an image is displayed in the snippet.

Here, the function *add speckle noise* is defined with three inputs. *image, gamma* and *sigma* represent the input image, the γ value and the standard deviation, respectively. The *gamma* and *sigma* parameters directly control the magnitude of the noise. The *np.random.randn* method generates a random noise matrix with the input image size. The multiplicative noise variable contains $I^{\gamma}(x, y)$ $\eta(x, y)$ matrix and it is added to the input image by element-wise addition, resulting a speckled image. The pixel values of the noisy image are clipped to ensure they fall within the valid range of 0–255. This step prevents any out-of-range values that may have occurred due to the noise addition. Finally, the speckled image is converted back to the unsigned 8-bit integer format with *np.uint*8 to ensure it has the appropriate data type for image representation (figure 10.1).

```python
import numpy as np
import matplotlib.pyplot as plt
import cv2


image1="C:/Users/Rishab Sarkar/Downloads/drive-download-20230510T180319Z-001/1.png"
image2="C:/Users/Rishab Sarkar/Downloads/drive-download-20230510T180319Z-001/2.png"
image3="C:/Users/Rishab Sarkar/Downloads/drive-download-20230510T180319Z-001/3.png"
image4="C:/Users/Rishab Sarkar/Downloads/drive-download-20230510T180319Z-001/4.png"

img1= cv2.imread(image1, 0)
img2= cv2.imread(image2, 0)
img3= cv2.imread(image3, 0)
img4= cv2.imread(image4, 0)
#image reshape
img1=cv2.resize(img1, (300,200))
img2=cv2.resize(img2, (300,200))
img3=cv2.resize(img3, (300,200))
img4=cv2.resize(img4, (300,200))
```

First, we have imported the necessary libraries to read and resize the images.

Figure 10.1. Test images used for qualitative and quantitative study. (a) Original clean image (GT). (b) Speckled image (Noisy).

```
#plotting clean image
plt.subplot(2,2,1)
plt.imshow(img1,cmap='gray')
plt.subplot(2,2,2)
plt.imshow(img2,cmap='gray')
plt.subplot(2,2,3)
plt.imshow(img3,cmap='gray')
plt.subplot(2,2,4)
plt.imshow(img4,cmap='gray')
```

After that, we plotted the clean images. Next, we introduced speckle noise to these images and plotted the speckled images.

```
#plotting clean image
plt.subplot(2,2,1)
plt.imshow(noisy_image1,cmap='gray')
plt.subplot(2,2,2)
plt.imshow(noisy_image2,cmap='gray')
plt.subplot(2,2,3)
plt.imshow(noisy_image3,cmap='gray')
plt.subplot(2,2,4)
plt.imshow(noisy_image4,cmap='gray')
```

10.3 Speckle noise reduction methods

Methods for reducing speckle noise, commonly referred to as despeckling methods, are used to diminish its effects and improve the visual appeal and diagnostic accuracy of images. The despeckling techniques can be broadly classified into conventional despeckling methods and data-driven approaches.

10.3.1 Conventional despeckling methods

Conventional despeckling methods include filters that operate either in the spatial domain or transform domain. Spatial domain despeckling methods are computationally efficient, operating directly on the image raw pixel values. These techniques assume that noise in the image is correlated with the underlying signal. Median filtering [23], mean filtering [7], and adaptive filtering like Lee [42], Frost [5], and Kuan filters [21] are examples of spatial domain despeckling methods. Although the aforementioned filters suppress speckle noise, they are ineffective in the presence of high speckle noise because the image details such as edges are blurred [36].

On the other hand, transformed domain approaches operate on the transformed coefficients of the image. These techniques take advantage of the fact that the noise in the image may have a different statistical distribution than the underlying signal in a transformed domain. Algorithms developed in the wavelet domain [28], discrete cosine transform (DCT) [1], or principal component analysis (PCA) [4] are examples of transform domain despeckling techniques. The transform domain despeckling methods are comparatively better for reducing the speckle noise and for preserving the details of the image. However, the filter implementation in the transform domain is computationally intensive and it can even introduce blocking artifacts in the output images [32]. In literature, some researchers have attempted to combine the goodness of both spatial and transform domain methods and proposed hybrid methods to achieve further improvement in the despeckling performance [12].

The Python code snippets for implementing different popular conventional despeckling methods are provided.

10.3.1.1 Lee filter
The following Python function defines the Lee filter and the filtered output can be generated by calling it with the speckle noise corrupted image as one of the inputs to the function call. The function details are given below:

```python
from scipy.ndimage.filters import uniform_filter
from scipy.ndimage.measurements import variance

def lee_filter(image, window_size):
    image_mean = uniform_filter(image, (window_size, window_size))
    image_mean_square = uniform_filter(image**2, (window_size, window_size))
    image_variance = image_mean_square - image_mean**2

    overall_image_variance = variance(image)

    image_weights = image_variance / (image_variance + overall_image_variance)
    image_output = image_mean + image_weights * (image - image_mean)
    return image_output

# Apply Lee filter with window size of 3x3
filtered_image1 = lee_filter(noisy_image1, window_size=3)
filtered_image2 = lee_filter(noisy_image2, window_size=3)
filtered_image3 = lee_filter(noisy_image3, window_size=3)
filtered_image4 = lee_filter(noisy_image4, window_size=3)

#plotting filtered image
plt.subplot(2,2,1)
plt.imshow(filtered_image1,cmap='gray')
plt.title("img1")
plt.subplot(2,2,2)
plt.imshow(filtered_image2,cmap='gray')
plt.title("img2")
plt.subplot(2,2,3)
plt.imshow(filtered_image3,cmap='gray')
plt.title("img3")
plt.subplot(2,2,4)
plt.imshow(filtered_image4,cmap='gray')
plt.title("img4")
```

The Lee filter function takes an *input image* and a *window size* as parameters. The *window size* represents the dimensions of the filtering window. The function uses this *window size* to compute the mean of the input image, referred to as 'image mean', and the squared mean of the image, known as '*image_mean_square*'. The image variance, which is a measure of the variation in pixel values, is calculated by subtracting the squared mean from the mean squared. To determine the weights for the filtered image, the image variance is divided by the sum of the image variance and the overall variance of the image. The overall variance of the image is computed using the variance function from scipy.ndimage.measurements on the original noisy image. The filtered image is then generated by adding the mean of the image to the element-wise multiplication of the weights and the difference between the original image and its mean. This process is applied to multiple noisy images with a window size of 3 × 3, and the filtered results are stored in separate variables.

10.3.1.2 Kuan filter
The Python code for the Kuan filter is given by:

```python
def kuan_filter(image, window_size):
    # Convert image to float
    image = image.astype(np.float64)
    # Estimate local mean using uniform filter
    mean = uniform_filter(image, window_size)
    # Calculate the adaptive weight
    weight = np.where(image > mean, 1.0, image / mean)
    # Apply Kuan filter
    filtered_image = mean * weight
    return filtered_image.astype(np.uint8)


# Apply Kuan filter with window size of 3x3
filtered_image1 = kuan_filter(noisy_image1, window_size=3)
filtered_image2 = kuan_filter(noisy_image2, window_size=3)
filtered_image3 = kuan_filter(noisy_image3, window_size=3)
filtered_image4 = kuan_filter(noisy_image4, window_size=3)
```

The Kuan filter function takes an *input image* and a *window size* as parameters. The image is first converted to a floating-point data type for accurate calculations. The local mean of the image is estimated using a uniform filter, which computes the mean value for each pixel based on the specified window size. The computation of the adaptive weight is done by comparing each pixel of the image with its corresponding mean value. If a pixel value is greater than the mean, it is assigned a weight of 1.0. Otherwise, the weight is calculated as the pixel value divided by the mean. This step determines the contribution of each pixel to the filtered result. The Kuan filter is then applied by multiplying the mean image by the weight image element-wise. This amplifies pixels with higher weights while attenuating those with lower weights, resulting in a filtered image. Finally, the filtered image is converted back to an unsigned 8-bit integer data type, which helps conserve memory and ensures compatibility with other image processing operations that specifically require 8-bit images.

10.3.1.3 SBF filter
The Python code for squeeze box filter is given by:

```python
def sbf_filter(image, sigma_s, sigma_r):
    # Convert image to float
    image = image.astype(np.float64)
    # Normalizing the image to the range [0, 1]
    image = image / 255.0

    # Create a grid of coordinates
    rows, cols = image.shape
    X, Y = np.meshgrid(np.arange(cols), np.arange(rows))

    # Initialize the filtered image
    filtered_image = np.zeros_like(image)

    # Iterate over each pixel in the image
    for i in range(rows):
        for j in range(cols):
            # Get the center pixel intensity
            center = image[i, j]

            # Calculate the spatial distance weights
            spatial_weights = np.exp(-((X - j)**2 + (Y - i)**2) / (2 * sigma_s**2))

            # Calculate the range distance weights
            range_weights = np.exp(-(image - center)**2 / (2 * sigma_r**2))

            # Calculate the combined weights
            weights = spatial_weights * range_weights

            # Calculate the filtered pixel value
            filtered_pixel = np.sum(weights * image) / np.sum(weights)

            # Update the filtered image
            filtered_image[i, j] = filtered_pixel

    # Convert the filtered image to unsigned 8-bit integer format
    filtered_image = (filtered_image * 255).astype(np.uint8)

    return filtered_image
    # Apply SBF filter with sigma_s = 3 and sigma_r = 0.1
    filtered_image1 = sbf_filter(noisy_image1, sigma_s=3, sigma_r=0.1)
    filtered_image2 = sbf_filter(noisy_image2, sigma_s=3, sigma_r=0.1)
    filtered_image3 = sbf_filter(noisy_image3, sigma_s=3, sigma_r=0.1)
    filtered_image4 = sbf_filter(noisy_image4, sigma_s=3, sigma_r=0.1)
```

The SBF filter function requires three parameters: the input image, the spatial standard deviation (sigma_s), and the range standard deviation (sigma_r). The function performs various steps to filter the image and produce the output. First, the input image is converted to a floating-point data type for precise calculations and then normalized to the range of [0, 1] by dividing it by 255, assuming the input image is in the range of [0, 255]. Next, a grid of coordinates representing the spatial coordinates of each pixel in the image is created using the np.meshgrid function. An empty array, called the 'filtered image,' is created to store the resulting filtered image, having the same shape as the input image. The code then iterates over each pixel in the image using nested loops. For each pixel, the center intensity value is obtained. Spatial distance weights, known as spatial weights, are calculated using the Euclidean distance between each pixel and the center pixel. The np.exp function is used to compute the exponential decay of the distances. Similarly, range distance weights, called range weights, are calculated based on the difference between each pixel and the center intensity value, using the np.exp function for exponential decay. The spatial and range distance weights are multiplied together to obtain the combined weights. The filtered pixel value is computed as the weighted sum of the image pixels divided by the sum of the weights. The resulting filtered pixel value is assigned to the corresponding location in the filtered image array. After processing all pixels, the filtered image is rescaled back to the range of [0, 255] by multiplying by 255 and converting it to the unsigned 8-bit integer format. Finally, the filtered image is returned as the output of the function.

10.3.1.4 *SRAD filter*
The Python code for the SRAD filter is given as:

```python
def srad_filter(image, num_iterations, delta_t, lambda_value):
    # Convert image to float
    image = image.astype(np.float64)

    # Perform SRAD filtering
    for _ in range(num_iterations):
        # Compute gradient of the image
        gradient_x = cv2.Sobel(image, cv2.CV_64F, 1, 0, ksize=3)
        gradient_y = cv2.Sobel(image, cv2.CV_64F, 0, 1, ksize=3)

        # Compute diffusivity function
        diffusivity = 1 / (1 + (gradient_x**2 + gradient_y**2) / lambda_value**2)

        # Update image using anisotropic diffusion equation
        image += delta_t * (diffusivity * (gradient_x**2 + gradient_y**2) -
                    (1 - diffusivity) * image)

        # Normalize the filtered image
        min_val = np.min(image)
        max_val = np.max(image)
        filtered_image = ((image - min_val) / (max_val - min_val) * 255).astype(np.uint8)

    return filtered_image
```

```
# Apply SRAD filter with 10 iterations, delta_t = 0.1, and lambda = 0.25
filtered_image1 = srad_filter(noisy_image1, num_iterations=10, delta_t=0.1, lambda_value=0.25)
filtered_image2 = srad_filter(noisy_image2, num_iterations=10, delta_t=0.1, lambda_value=0.25)
filtered_image3 = srad_filter(noisy_image3, num_iterations=10, delta_t=0.1, lambda_value=0.25)
filtered_image4 = srad_filter(noisy_image4, num_iterations=10, delta_t=0.1, lambda_value=0.25)
```

The SRAD filter function takes four parameters: the input image, the number of iterations for SRAD filtering, the time step (delta_t), and a parameter controlling the diffusivity function (lambda value). The function performs several steps to apply anisotropic diffusion to the image and obtain the filtered result. First, the input image is converted to a floating-point data type for precise calculations using the astype function from the NumPy library. Within each iteration, the gradient of the image is computed using the cv2.Sobel function from the OpenCV library. This calculates the derivative of the image in the x and y-directions using the Sobel operator. The diffusivity function is then computed using the gradient information. It is calculated as $1/(1 + (\text{gradient } x2 + \text{gradient } y2)/\text{lambda value}^{**}2)$, where gradient x and gradient y represent the gradients in the x and y-directions, respectively. The image is updated using the anisotropic diffusion equation, which enhances or smooths the image based on the diffusivity value at each pixel location. The updated image is calculated as image += delta t * (diffusivity * (gradient x2 + gradient y2)−(1−diffusivity) * image). After completing the specified number of iterations, the filtered image is obtained. To ensure the image is within the range of [0, 255], it undergoes normalization by determining the minimum and maximum values (min_val and max_val) present in the image. The normalized image is then calculated using the formula ((image−min_val)/(max_val −min_val) * 255). Finally, the filtered image is converted to the unsigned 8-bit integer format using the astype function and returned as the output of the function.

10.3.1.5 Frost filter
The Python code for the Frost filter is given as:

```
def frost_filter(image, window_size, alpha):
    # Convert image to float
    image = image.astype(np.float64)

    # Calculate the local mean
    local_mean = cv2.blur(image, (window_size, window_size))

    # Calculate the local variance
    local_variance = cv2.blur(image**2, (window_size, window_size)) - local_mean**2

    # Calculate the filtered image using Frost filter equation
    filtered_image = image - alpha * (image - local_mean) / np.maximum(local_variance, 1e-8)

    # Normalize the filtered image
    filtered_image = cv2.normalize(filtered_image, None, 0, 255, cv2.NORM_MINMAX, dtype=cv2.CV_8U)

    return filtered_image

# Apply Frost filter with window size = 3 and alpha = 1.5
filtered_image1 = frost_filter(noisy_image1, window_size=3, alpha=1.5)
filtered_image2 = frost_filter(noisy_image2, window_size=3, alpha=1.5)
filtered_image3 = frost_filter(noisy_image3, window_size=3, alpha=1.5)
filtered_image4 = frost_filter(noisy_image4, window_size=3, alpha=1.5)
```

The Frost filter function requires three parameters: the input image, the window size for local processing, and the alpha parameter controlling the filtering strength. The function performs several steps to apply the Frost filter and obtain the filtered image. First, the input image is converted to a floating-point data type (np.float64) using the astype function from the NumPy library for accurate calculations. Next, the local mean is calculated using the cv2.blur function from OpenCV, which applies a box filter to determine the mean value for each pixel within a local window of size (window size, window size). The local variance is computed by squaring the image using the ** operator and then applying the cv2.blur function to calculate the mean of the squared image. The local variance is obtained by subtracting the square of the local mean from the calculated mean of the squared image. The filtered image is obtained using the Frost filter equation: filtered image=image−alpha * (image −local mean)/np.maximum(local variance, 1e−8). This equation adjusts the pixel intensities based on the difference between the pixel value and its local mean, divided by the local variance. The np.maximum function is used to avoid division by zero, ensuring that the denominator is at least 1e-8. Subsequently, the filtered image is normalized to the range [0, 255] using the cv2.normalize function from OpenCV. This function maps the pixel values of the filtered image from their current range to the range [0, 255] while preserving the relative intensity distribution. Finally, the normalized filtered image is returned as the output of the function.

To evaluate the effectiveness of these approaches in improving image quality, performance assessment metrics are employed [37]. These metrics serve as objective measures to assess the performance of different despeckling algorithms and enable researchers to compare their effectiveness for specific applications.

A few popular performance metrics for despeckling algorithms are discussed below. Signal-to-noise ratio (SNR) measures the quality of an image by comparing the original image x and despeckled image y. A higher SNR value indicates better image quality.

$$SNR(x, y) = \frac{10\sum_{x=1}^{M}\sum_{y=1}^{N}(x)^2}{\sum_{x=1}^{M}\sum_{y=1}^{N}(x-y)^2} \qquad (10.3)$$

where M and N are the dimensions of the image.

Structural similarity index metric (SSIM) is another measure used to evaluate the visual similarity between the original and despeckled images. Luminance, contrast, and structure are the three basic elements of human visual perception that are taken into consideration by the SSIM index. The mathematical expression for SSIM is given in equation (10.4).

$$SSIM(x, y) = \frac{(2\mu_x\mu_y + C_1)(2\sigma_{xy} + C_2)}{\left(\mu_x^2 + \mu_y^2 + C_1\right)\left(\sigma_x^2 + \sigma_y^2 + C_2\right)} \qquad (10.4)$$

where μ_x and μ_y are the means of x and y, σ_x and σ_y are the standard deviations of x and y, σ_{xy} is the covariance between x and y, and c_1 and c_2 are constants to stabilize the division in case of weak denominator.

Edge preserving index (EPI) measures the degree to which a despeckling algorithm preserves the edges in an image.

$$EPI = \frac{\sum_{x=1}^{N} w_x \cdot \left(1 - \frac{|\nabla I_x|}{|\nabla I|}\right)}{\sum_{x=1}^{N} \omega_x} \tag{10.5}$$

where N is the total number of pixels in the image, I_x is the intensity value of the xth pixel, ∇I_x is the gradient magnitude of the xth pixel, $|\nabla I|$ is the average gradient magnitude of the image, and w_x is a weighting function that gives more weight to pixels near the edges. Contrast-to-noise ratio (CNR) measures the image quality that reflects the ability of a despeckling algorithm to preserve contrast between different regions of an image while reducing noise.

$$CNR = \frac{|m_1 - m_2|}{\sqrt[2]{\sigma_1^2 + \sigma_2^2}} \tag{10.6}$$

where m_1 and m_2 are the mean pixel values of two different regions of interest (ROIs) in the image, and σ_1 and σ_2 are the corresponding standard deviations of the pixel values within the ROIs.

Equivalent number of looks (ENL) measures the degree to which a despeckling algorithm has reduced noise in the image while preserving the spatial resolution.

$$ENL = \frac{\mu^2}{\sigma^2} \tag{10.7}$$

where μ is the mean value and σ is the standard deviation of the image.

Now to calculate the parametric measures of the filtered image we have defined the function using Python, which is given as:

```python
def psnr(original, reconstructed):
    mse = np.mean((original - reconstructed) ** 2)
    max_pixel = np.max(original)
    return 20 * np.log10(max_pixel / np.sqrt(mse))


from scipy.ndimage import sobel
def epi(original, reconstructed):
    gradient_magnitude = np.abs(sobel(original))
    average_gradient_magnitude = np.mean(gradient_magnitude)
    weights = gradient_magnitude / average_gradient_magnitude
    epi_value = np.sum(weights * (1 - np.abs(sobel(reconstructed)) / average_gradient_magnitude)) /
np.sum(weights)
    return epi_value


def cnr(original, reconstructed):
    mean_original = np.mean(original)
    std_original = np.std(original)
    mean_reconstructed = np.mean(reconstructed)
```

```
    std_reconstructed = np.std(reconstructed)
    return abs(mean_original - mean_reconstructed) /
np.sqrt((std_original ** 2 + std_reconstructed ** 2) / 2)

def enl(image):
    mean = np.mean(image)
    variance = np.var(image)
    return (mean ** 2) / variance

# Load sample images
original_image = img1

# Load reconstructed image and resize to match shape of original image
reconstructed_image = filtered_image1
reconstructed_image = cv2.resize(reconstructed_image, (original_image.shape[1],
original_image.shape[0]))

# Call PSNR function
psnr_value = psnr(original_image, reconstructed_image)
print("PSNR value:", psnr_value)

# Call EPI function
epi_value = epi(original_image, reconstructed_image)
print("EPI value:", epi_value)

# Call CNR function
cnr_value = cnr(original_image, reconstructed_image)
print("CNR value:", cnr_value)

# Call ENL function
enl_value = enl(original_image)
print("ENL value:", enl_value)

mse = np.mean((original_image - reconstructed_image) ** 2)
print("MSE value:",mse)
```

Please note that the outputs obtained with the different aforementioned filters are provided in the next section on experimental results.

10.3.2 Data-driven despeckling approaches

The data-driven approaches are becoming increasingly popular due to their superior performance to the conventional despeckling methods. In data-driven approaches, deep learning (DL) models are used for despeckling which rely on large amounts of data to learn a model for removing speckle noise from an image. These DL models typically require a large amount of training data to learn the complex relationships between input image and output despeckled image. Once the deep learning model is trained, it can be used to despeckle new images. These methods can be very effective at removing noise, particularly when the noise is complex and difficult to model using traditional filtering techniques. These data-driven approaches are more superior than conventional approaches because of their ability to learn from data, which includes learning patterns and relationships in the data, flexibility to specific problems and datasets, robustness, scalability and automation [2].

The steps involved in the design cycle of the data-driven approaches are briefly described next. The first step is to gather and prepare the data for training. This may involve collecting a diverse and representative dataset, cleaning and preprocessing the data, and splitting it into training, testing, and validation sets. In the model architecture design stage, we have to define the architecture of our deep learning model. This includes choosing the type of neural network and designing its structure. After the model architecture is defined, the model is trained using the training dataset. During training, the model learns to make predictions by adjusting its internal parameters like weights and biases based on the input data and the desired outputs. This typically involves an iterative optimization process using algorithms like backpropagation and gradient descent.

After training, it is crucial to assess the performance and generalization ability of the model on unseen data. The validation dataset, which is separate from the training set, is used for this purpose. The model's performance on the validation set helps evaluate its ability to generalize to new, unseen data and identify potential issues like overfitting. After the model has been trained and validated, it is tested using a separate testing dataset. This final evaluation provides an unbiased assessment of the model's performance on unseen data and helps estimate its real-world performance. Based on the results of the model evaluation, you may need to fine-tune and optimize the model further. This can involve adjusting hyperparameters, modifying the model architecture, or applying regularization techniques to improve performance and reduce overfitting. Then, it can be deployed for real-world use. This involves integrating the model into a production environment or application, where it can make predictions on new, unseen data.

In recent years, one of the most promising approaches for despeckling medical images is DL, which utilizes artificial neural networks to learn from large datasets and make predictions on new data. Supervised learning in DL involves training a neural network on a dataset of both noisy and clean images to learn the relationship between them and denoise new images. Convolutional neural networks (CNNs) [30] and autoencoder networks [37] are examples of supervised DL algorithms used for despeckling. In contrast, unsupervised learning in DL involves training a neural network on a dataset of noisy images only to extract features and patterns without reference to clean images. Generative adversarial networks (GANs) [41] are examples of unsupervised DL algorithms used for despeckling. DL offers several advantages, including learning complex patterns and features in medical images, training on large datasets for improved robustness and generalizability, and processing images in real time for applications requiring rapid decision making.

Despite the potential benefits of DL for despeckling medical images, there are still several challenges that need to be addressed. One of the major challenges is the availability of annotated datasets for training the DL algorithms. Annotated datasets are essential for training DL algorithms as they provide ground truth information for the algorithm to learn from. However, annotated medical image datasets are often limited due to privacy concerns and the difficulty of acquiring large amounts of annotated data. Another challenge is the interpretability of DL algorithms. DL algorithms can learn complex patterns and features in medical images, but it can be difficult to understand how the algorithm arrived at its predictions. This can be a significant issue for medical applications where the ability to explain and understand the algorithm's decisions is critical.

The Python code of such DL-based models are as follows:

10.3.2.1 Convolutional autoencoder

Here we have used Dataset_BUSI_with_GT from Kaggle to train our convolutional autoencoder architecture.

```
import re
import numpy as np
import pandas as pd
import random
import math
import cv2
import seaborn as sns
from glob import glob
import os
import itertools
import tensorflow as tf
from tensorflow.keras.preprocessing.image import ImageDataGenerator
from tensorflow.keras.layers import Dense, Conv2D, MaxPooling2D,MaxPool2D
from tensorflow.keras.layers import UpSampling2D, Flatten,
Input,LeakyReLU,BatchNormalization,Dropout
from tensorflow.nn import atrous_conv2d
from tensorflow.keras.models import Model, Sequential
from tensorflow.keras.optimizers import Adam
from tensorflow.keras.callbacks import EarlyStopping, ModelCheckpoint, ReduceLROnPlateau, CSVLogger
import matplotlib.pyplot as plt
from skimage.util import random_noise
from skimage.metrics import peak_signal_noise_ratio as psnr,mean_squared_error as mse
```

In order to build the architecture of the model, we import a range of Python packages and libraries that provide essential functionalities. These packages include popular deep learning frameworks such as TensorFlow, as well as relevant libraries for image processing and evaluation.

```
img_path = 'C:/Users/Rishab Sarkar/Downloads/archive (1)/Dataset_BUSI_with_GT/*/*.png'
images = glob(img_path)
images.sort()
```

Here, we define the path of the directory containing the ultrasound dataset. The glob function from the glob package is then used to extract the file paths of all PNG images in the directory and its subdirectories, using the * wildcard to match any characters in the file path. The extracted file paths are then sorted using the sort method of the resulting list.

```python
            #creating dictionary of images and masks
        image = []
        mask = []
        count = 1
        i= 0
        num = len(images)
        images_per_class = {}
        while i < num-1:
            print(str(count) + "/" + str(num), end="\r")
            img = images[i]


label = img.split('/')[-2]
#get label from image name
image_name = img.split('/')[-1].split(')')[0]
#get image name
if label not in images_per_class.keys():
  images_per_class[label] = {'image':[],'mask':[]}

  #make nested dictionary for each label
if img.split('/')[-1][-5] == ')': #get the image
  image = cv2.resize(cv2.imread(img,cv2.IMREAD_GRAYSCALE),(128,128))
  #read image
  images_per_class[label]['image'].append(image)
  mylist = images[i:]
  r = re.compile(r".*"+re.escape(image_name)+r"\)_mask*.")
  #look for mask
  masklist = list(filter(r.match, mylist))
  if len(masklist) == 1:
    mask = cv2.resize(cv2.imread(masklist[0]),(128,128))
    images_per_class[label]['mask'].append(mask)
  else:  #there are two masks
    mask_1 = cv2.imread(masklist[0])
    mask_2 = cv2.imread(masklist[1])
    mask = cv2.resize(cv2.bitwise_or(mask_1,mask_2),(128,128))
    images_per_class[label]['mask'].append(mask)
i+=1

count += 1
```

In this section, the code begins by initializing an empty dictionary that will be used to store images and masks. It then reads a set of images from a specified file path using a loop. For each image, the code resizes it to a size of 128 × 128 pixels and stores it in the dictionary. The label for each image is extracted from its file path and used as a key in the dictionary. Next, it searches for the corresponding mask for each image in the remaining list of images. If only one mask is found, it is resized to 128 × 128 pixels and stored in the dictionary. In the case of two masks, they are combined using the bitwise OR operation to create a single mask. This combined mask is then resized to 128 × 128 pixels and stored in the dictionary. Finally, the code returns the dictionary containing the resized images and masks, allowing for further processing and analysis.

```python
#number of images per class
labels = []
for key,value in images_per_class.items():
    for k,v in value.items():
        labels.extend([k]*len(v))
        print("{0} : {1} : {2} ".format(key,k, len(v)))

#dataset for despeckling consists of only images
data = images_per_class['archive (1)']['image']

#normalizing pixel values in the dataset
data = [img/255 for img in data]
random.shuffle(data)

#Plotting set of images
def plot_images(data):
    fig,ax=plt.subplots(1,5)
    fig.set_size_inches(40,20)
    for i in range(5,10):
        ax[i-5].imshow(data[i], cmap='gray')
    plt.show()

def add_speckle_noise(image, gamma, sigma):
    # Ensure the input image is a numpy array
    image = np.array(image, dtype=float)
    # Generate zero-mean Gaussian noise
    noise = np.random.normal(0, sigma, size=image.shape)
    # Apply speckle noise according to the given mathematical model
    noisy_image = image + (image ** gamma) * noise
    # Ensure pixel values are in the valid range [0, 255]
    noisy_image = np.clip(noisy_image, 0, 255)
    return noisy_image.astype(np.uint8)
```

```
speckled_data=[]
for img in data:
    speckled=add_speckle_noise(img, gamma=0.5, sigma=30)
    #introducing speckle noise
    speckled_data.append(speckled)

speckled_data=np.array(speckled_data)
```

Then the code begins by printing the number of images per class in the dataset, providing an overview of the dataset distribution. Next, it creates a denoising dataset containing only the images and normalizes the pixel values between 0 and 1, ensuring consistent data representation for further processing. Afterwards, the dataset is shuffled to introduce randomness, and a set of images from the shuffled dataset is plotted for visualization purposes. Following that, the code applies speckle noise to each image in the dataset, simulating a noisy environment. The resulting noisy images are stored in a new array called 'noised dataset,' which contains the same images as the original dataset but with added noise. Finally, the 'noised dataset' array is converted to a Numpy array, providing a convenient format for subsequent operations and analysis on the noisy images.

```
#Splitting dataset for training and testing
x_train = speckled_data[:600]
x_train_clean = data[:600]
x_test=speckled_data[600:]
x_test_clean = data[600:]
```

The code proceeds by splitting the dataset into training and testing sets. The speckled images are stored in a variable called 'speckled_dataset,' and the corresponding clean images are stored in a variable called 'dataset.' For the training set, the first 600 images from both the speckled dataset and the clean dataset are selected. The speckled images are stored in the variable 'x_train,' and the corresponding clean images are stored in the variable 'x_train_clean.' The remaining images in both datasets are designated for the testing set. The speckled images are stored in the variable 'x_test,' and the corresponding clean images are stored in the variable 'x_test_clean.' This split allows for distinct sets of images to be used for training and testing, facilitating the evaluation of the model's performance on unseen data.

```
#building model architecture
def speckle_autoencoder():
    i=Input(shape=(128,128,1))
    #encoder
    x = Conv2D(128, (3,3), activation='relu', padding='same')(i)
    x = MaxPooling2D((2,2), padding='same')(x)
```

```
x = Conv2D(128, (3,3), activation='relu', padding='same')(x)
x = MaxPooling2D((2,2), padding='same')(x)
x = Conv2D(128, (3,3), activation='relu', padding='same')(x)
x = Conv2D(128, (3,3), activation='relu', padding='same')(x)

#decoder
x = Conv2D(128, (3,3), activation='relu', padding='same')(x)
x = Conv2D(128, (3,3), activation='relu', padding='same')(x)
x = UpSampling2D((2,2))(x)
x = Conv2D(128, (3,3), activation='relu', padding='same')(x)
x = UpSampling2D((2,2))(x)
x = Conv2D(1, (3,3), activation='sigmoid', padding='same')(x)

#model
autoencoder = Model(inputs=i, outputs=x)
autoencoder.compile(optimizer='Adam', loss='mse')
autoencoder.summary()

return autoencoder

autoencoder = speckle_autoencoder()
```

The code implements a function called 'speckle_autoencoder' using the Keras library. This autoencoder is specifically designed to take grayscale images with a size of 128×128 pixels as input and generate despeckled versions of the input images as output. The architecture of the autoencoder consists of two main parts: an encoder and a decoder. The encoder utilizes three convolutional layers with 128 filters each, followed by max-pooling layers, and two additional convolutional layers. This sequence of layers is responsible for extracting relevant features from the input images. On the other hand, the decoder part of the network is symmetric to the encoder. It includes two more convolutional layers with 128 filters each, followed by upsampling layers, and a final convolutional layer that generates the output, which is a denoised image. The 'speckle_autoencoder' function returns the constructed autoencoder model and compiles it using the Adam optimizer and mean squared error (MSE) loss, which is commonly used for image reconstruction tasks. Finally, the function assigns the autoencoder model to the variable 'autoencoder.' This implementation provides a ready-to-use autoencoder model that can be trained on speckled images to learn to reconstruct clean, despeckled versions of the input images.

```
epoch_num = 300
batch_num = 64

learning_rate = ReduceLROnPlateau(monitor='val_loss',
                        patience=8,
                        verbose=1,
                        factor=0.8,
                        min_lr=1e-10)
r = autoencoder.fit(x_train, x_train,
            epochs=epoch_num,
            batch_size=batch_num,
            shuffle=True,
            validation_data=(x_test, x_test),
            callbacks=[learning_rate],verbose=0)
#model evaluation
print("Train score:", autoencoder.evaluate(x_train,x_train))
print("Test score:", autoencoder.evaluate(x_test,x_test))
```

The training of the autoencoder model is performed using the training and testing datasets that were previously split. The code specifies the number of epochs and batch size for the training process. Additionally, it applies the ReduceLROnPlateau callback, which reduces the learning rate when a certain condition is met, helping to improve the model's performance. The model is then trained using the specified number of epochs and batch size. During training, the model learns to reconstruct despeckled images from the speckled input images. After training, the model is evaluated on both the training and testing datasets to assess its performance. The evaluation scores, which can include metrics such as accuracy or loss, are then displayed. This process allows for the training and evaluation of the autoencoder model, enabling the assessment of its ability to effectively despeckle images.

```
epoch_num = 300
batch_num = 64

learning_rate = ReduceLROnPlateau(monitor='val_loss',
                        patience=8,
                        verbose=1,
                        factor=0.8,
                        min_lr=1e-10)
r = autoencoder.fit(x_train, x_train,
            epochs=epoch_num,
            batch_size=batch_num,
            shuffle=True,
            validation_data=(x_test, x_test),
            callbacks=[learning_rate],verbose=0)
#model evaluation
print("Train score:", autoencoder.evaluate(x_train,x_train))
print("Test score:", autoencoder.evaluate(x_test,x_test))
```

Here the code defines a function named 'plotLearningCurve' that takes three arguments: 'history,' 'epoch num,' and 'batch num.' The 'history' argument represents the object returned by the fit() method during the model training process. It contains information about the training process, such as the loss and validation loss at each epoch. The 'epoch_num' parameter indicates the total number of epochs, while 'batch_num' represents the batch size used during training. Within the function, Matplotlib is utilized to create a plot that displays the training loss and validation loss for each epoch. The x-axis corresponds to the epoch number, and the y-axis represents the loss. The title of the plot includes the values of 'epoch_num' and 'batch_num' to provide context. Finally, the plot is displayed using plt.show(), enabling visualization of the learning curve that illustrates how the loss changes over the training process. This function facilitates the analysis of the model's training performance and the identification of potential issues such as overfitting or underfitting.

```
# Load the images and preprocess them
image1 = "C:/Users/Rishab Sarkar/Downloads/drive-download-20230510T180319Z-001/1.png"
image2 = "C:/Users/Rishab Sarkar/Downloads/drive-download-20230510T180319Z-001/2.png"
image3 = "C:/Users/Rishab Sarkar/Downloads/drive-download-20230510T180319Z-001/3.png"
image4 = "C:/Users/Rishab Sarkar/Downloads/drive-download-20230510T180319Z-001/4.png"
```

As the image pixel is in the range [0,255] so before predicting on the image, we have to normalize the image in the range [0,1] after this step

```
img1 = img1/255
img1 = img2/255
img1 = img3/255
img1 = img4/255

# Preprocess the images and reshape to match the input shape of the model
img1 = cv2.resize(img1, (128, 128))
img1 = img1.reshape(1, 128, 128, 1)

img2 = cv2.resize(img2, (128, 128))
img2 = img2.reshape(1, 128, 128, 1)

img3 = cv2.resize(img3, (128, 128))
img3 = img3.reshape(1, 128, 128, 1)

img4 = cv2.resize(img4, (128, 128))
img4 = img4.reshape(1, 128, 128, 1)

# Use the preprocessed images for prediction
pred1 = autoencoder.predict(img1)
pred2 = autoencoder.predict(img2)
pred3 = autoencoder.predict(img3)
pred4 = autoencoder.predict(img4)
```

In this portion the code involves the specification of file paths for four images (image1, image2, image3, image4) that need to be loaded and preprocessed. Each image is loaded using the cv2.imread() function with a grayscale flag (0) to load it as a grayscale image. The resulting images are stored in variables (img1, img2, img3, img4). To ensure compatibility with the model's input shape, the images are preprocessed by resizing them to a size of 128 × 128 pixels using the cv2.resize() function. Furthermore, the shape of each image is modified using the reshape() function to have dimensions of (1, 128, 128, 1). This additional dimension represents a batch of one image. Subsequently, the preprocessed images are utilized for prediction by invoking the autoencoder.predict() function on each image individually. The resulting predictions are then stored in variables (pred1, pred2, pred3, pred4). This process involves loading, resizing, reshaping, and predicting on the specified images, ultimately generating the corresponding predictions using the trained autoencoder model.

As the predicted images are in the range [0,1], we have to make it in original range i.e [0,255] after this step

```
# Reshape the predictions back to 2D images
pred1 = pred1*255
pred1 = pred1.astype(np.uint8)
pred2 = pred2*255
pred2 = pred2.astype(np.uint8)
pred3 = pred3*255
pred3 = pred3.astype(np.uint8)
pred4 = pred4*255
pred4 = pred4.astype(np.uint8)


# Plot the predictions
fig, axes = plt.subplots(2, 2, figsize=(10, 10))

axes[0, 0].imshow(pred1, cmap='gray')
axes[0, 0].set_title('Prediction 1')

axes[0, 1].imshow(pred2, cmap='gray')
axes[0, 1].set_title('Prediction 2')

axes[1, 0].imshow(pred3, cmap='gray')
axes[1, 0].set_title('Prediction 3')

axes[1, 1].imshow(pred4, cmap='gray')
axes[1, 1].set_title('Prediction 4')

plt.tight_layout()
plt.show()
```

After reshaping the predictions pred1, pred2, pred3, and pred4 to a 2D shape of size 128×128 using the reshape function, a figure is created along with a 2×2 grid of subplots using the plt.subplots(2, 2, figsize = (10, 10)) function. This arrangement results in a figure containing four subplots arranged in a 2×2 grid. The figsize parameter determines the dimensions of the figure. For each subplot, the imshow function is used to display the corresponding reshaped prediction array. By setting the cmap='gray' argument, the colormap is set to grayscale, ensuring that the images are displayed in grayscale. The set_title function is utilized to set a title for each subplot. Finally, plt.tight_layout() ensures that the subplots are properly spaced, and plt.show() is called to display the plot with the predictions. This process results in a visualization of the reshaped predictions in a grid format.

Now we can calculate the parametric measures for it with the help of Python code given in the conventional approach section.

10.3.2.2 Noise2void

Here the algorithm learns despeckling from a single speckled image. The Python code for the algorithm is given as follows:

```python
import tensorflow as tf
import n2v
from n2v.models import N2VConfig, N2V
import numpy as np
from csbdeep.utils import plot_history
from n2v.utils.n2v_utils import manipulate_val_data
from n2v.internals.N2V_DataGenerator import N2V_DataGenerator
from matplotlib import pyplot as plt
import os
import cv2


# load data and extract patches for training and validation.
datagenerator = N2V_DataGenerator()
```

Here we imported all the necessary libraries required to use the noise2void model. After that, we created a 'datagenerator' variable to load the data.

```python
image1="C:/Users/Rishab Sarkar/Downloads/drive-download-20230510T180319Z-001/1.png"
img1= cv2.imread(image1, 0)

#image reshape
img1=cv2.resize(img1, (300,200))


def add_speckle_noise(image, gamma, sigma):
    # Generate random noise that matches the dimension of the image
    sigma_noise = np.random.randn(*image.shape) * sigma
    # Generate multiplicative noise with the same dimension of the image
    multiplicative_noise = np.random.randn(*image.shape) * gamma
    # Apply the generated multiplicative noise in the image
    speckle_image = image + image**multiplicative_noise + sigma_noise
    # Ensure the pixel values within the valid range of 0-255
    speckle_image = np.clip(speckle_image, 0, 255)
    # Convert the image back to unsigned 8-bit integer format
    speckle_image = speckle_image.astype(np.uint8)
    return speckle_image

# Add speckle noise with sigma = 30
noisy_image1 = add_speckle_noise(img1, gamma=0.5, sigma=30)

# Save the images
cv2.imwrite("resized_image1.png", noisy_image1)
```

Only a single image, i.e., img1 was taken to add speckle noise which will be used for training purposes.

```
# load the '.png' file from the directory and the function will return a list of images.
imgs = datagenerator.load_imgs_from_directory(directory = "C:/Users/Rishab
```

```
Sarkar/Desktop/cnn/grayscale_img/", filter='resized_image1.png', dims='YX')
```

```
# shape of the image
print('shape of loaded images: ',imgs[0].shape)
```

For training purposes, we have to take lossless format images. Here we have taken a ".png" format image and printed the shape of the loaded image.

```
patch_size = 64
# Patches are extracted from images and combined into a single numpy array
patch_shapes = (patch_size,patch_size)
patches = datagenerator.generate_patches_from_list(imgs, shape=patch_shapes)
```

```
# Utilizing non-overlapping patches allows us to divide them into training and validation sets.
train_val_split = int(patches.shape[0] * 0.8)
X_train = patches[:train_val_split]
X_valid = patches[train_val_split:]
print(X_train.shape)
print(X_train.dtype)
print(X_valid.shape)
```

In this portion the code defines the patch size as 64, indicating the desired dimensions of the square patches. A tuple called patch_shapes is created with the dimensions of the patches set to patch size x patch size. Using the generate_-patches_from_list method of the datagenerator object, the patches are generated from the list of images, imgs, and combined into a single Numpy array. The shape of the patches will be defined by patch_shapes. To split the patches into training and validation sets, the code calculates the train_val_split index by multiplying the total number of patches (patches.shape[0]) by 0.8 and converting it to an integer, indicating that 80% of the patches will be used for training. The first train_val_split patches are assigned to the X_train variable, which will be used for training. The remaining patches from the index train_val_split onwards are assigned to the X_valid variable, which will be used for validation. This ensures that the patches are appropriately divided for training and validation purposes.

```
# training is shown once per epoch.
train_batch = 32
config = N2VConfig(X_train, unet_kern_size=3,
         unet_n_first=64, unet_n_depth=3, train_steps_per_epoch=int(X_train.shape[0]/train_batch),
train_epochs=300, train_loss='mse',
         batch_norm=True, train_batch_size=train_batch, n2v_perc_pix=0.198,
n2v_patch_shape=(patch_size, patch_size),
         n2v_manipulator='uniform_withCP', n2v_neighborhood_radius=5,
single_net_per_channel=False)

# Parameters are stored within the config-object.
vars(config)
```

Then we trained the model for 30 epochs and the mean square error was taken as a loss function for it. Finally, the vars(config) function is used to retrieve a dictionary containing the attributes and their values of the config object. This allows printing all the configuration parameters as a dictionary.

```
# a name used to identify the model
name_model= 'n2v'
# The base directory where our model will be located
base_dir = 'models'
# We are currently in the process of creating our network model.
model = N2V(config, name_model, basedir=base_dir)

#start training
history = model.train(X_train, X_valid)
```

A base directory was created to save the model and the name of the model was given as 'n2v'.

```
# Plot training and validation loss
plt.plot(history.history['loss'])
plt.plot(history.history['val_loss'])
plt.title('Model Loss')
plt.ylabel('Loss')
plt.xlabel('Epoch')
plt.legend(['Training loss', 'Validation loss'], loc='upper right')
plt.show()
```

After that we plotted the training loss and the validation loss for each epoch.

```
# We import all our dependencies.
from n2v.models import N2V
import numpy as np
from matplotlib import pyplot as plt
from matplotlib.image import imread, imsave

# To load a previously trained model, create a new N2V-object without providing a 'config' parameter.
name_model = 'n2v'
base_dir = 'models'
model = N2V(config=None, name=name_model, basedir=base_dir)
```

```
# We load the data we want to process.
img1 = imread("C:/Users/Rishab Sarkar/Desktop/cnn/grayscale_img/resized_image1.png")
img2 = imread("C:/Users/Rishab Sarkar/Desktop/cnn/grayscale_img/resized_image2.png")
img3 = imread("C:/Users/Rishab Sarkar/Desktop/cnn/grayscale_img/resized_image3.png")
img4 = imread("C:/Users/Rishab Sarkar/Desktop/cnn/grayscale_img/resized_image4.png")


n2v_pred1 = model.predict(img1, axes='YX')
n2v_pred2 = model.predict(img2, axes='YX')
n2v_pred3 = model.predict(img3, axes='YX')
n2v_pred4 = model.predict(img4, axes='YX')
```

Now this code attempts to load the network weights from the file 'weights_best. h5', assuming that the weights file is present in the specified directory. Four grayscale images, img1, img2, img3, and img4, are loaded using the imread function from Matplotlib. The file paths of the images are provided as arguments to imread. To generate predictions for each input image, the predict method of the N2V model is called. The predictions are generated along the y and x-axes (2D prediction) by setting the axes parameter to 'YX'. The resulting predictions are stored in variables n2v_pred1, n2v_pred2, n2v_pred3, and n2v_pred4, respectively. These variables will contain the predicted outputs of the N2V model for the corresponding input images.

10.4 Experimental results

In this section, we evaluate the results qualitatively and quantitatively. One popular method of qualitative evaluation is the visual inspection of the results obtained by different methods. To evaluate the effectiveness of different techniques, figures 10.2 and 10.3 are the results obtained for different despecking methods at two different noise levels indicated by the standard deviation at $\sigma = 30$ and $\sigma = 40$, respectively. Also, the quantitative evaluation results at $\sigma = 30$ and $\sigma = 40$ are provided in tables 10.1 and 10.2, respectively. For that purpose, we presented the results obtained using the quality measures such as PSNR, EPI, CNR, ENL and MSE.

After conducting comprehensive qualitative and quantitative evaluations, it can be confidently stated that deep learning models consistently outperform conventional despeckling methods, delivering superior results. Moreover, DL models exhibit enhanced generalization capabilities, allowing them to perform well on a wide range of images, including those with different levels of noise and varying textures.

10.5 Conclusions

Speckle noise refers to the granular noise pattern that arises due to the coherent nature of the imaging systems such as medical US, OCT and SAR images. Despeckling methods dealt with the adverse effects of speckle noise, resulting in superior image quality, improved information extraction accuracy, streamlined quantitative analysis, and image content preservation. Different despeckling methods, including conventional methods and data-driven techniques, have been

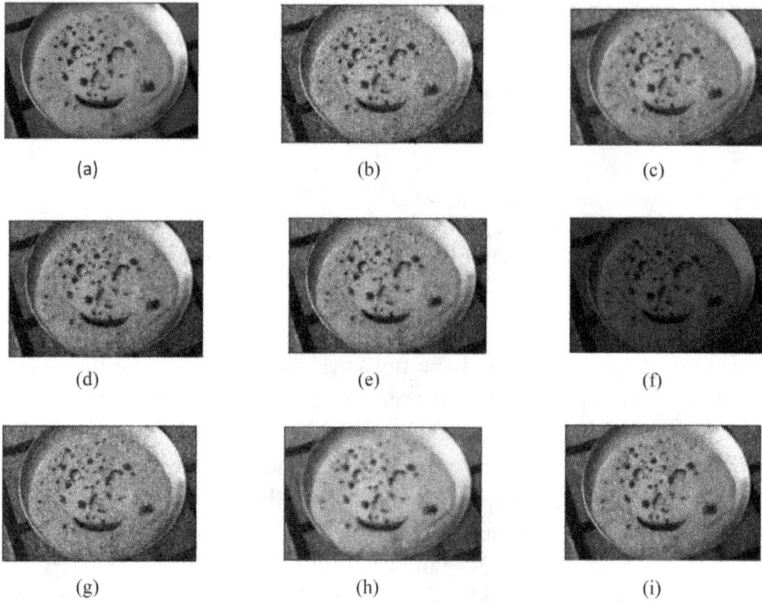

Figure 10.2. Results obtained with different filters applied to the given image. (a) Original image, (b) Speckled image ($\gamma = 0.5$, $\sigma = 30$), (c) Lee filter result, (d) Kuan filter result, (e) SBF filter result (f) SRAD filter result, (g) Frost filter result, (h) Noise2Void filter result, (i) Convolutional autoencoder filter result.

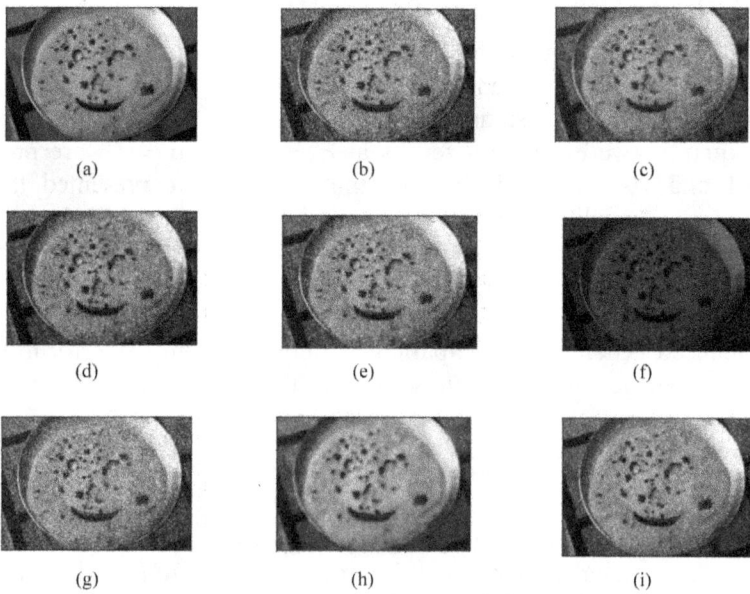

Figure 10.3. Results obtained with different filters applied to the given image. (a) Original image, (b) Speckled image ($\gamma = 0.5$, $\sigma = 40$), (c) Lee filter result, (d) Kuan filter result, (e) SBF filter result (f) SRAD filter result, (g) Frost filter result, (h) Noise2Void filter result, (i) Convolutional autoencoder filter result.

Table 10.1. Quantitative result on speckled image corrupted with $\sigma = 30$ value.

Method	PSNR	EPI	CNR	ENL	MSE
Lee [42]	22.2622	0.5555	0.2258	3.7240	386.2402
Kuan [21]	28.7873	0.0593	0.1170	3.7240	85.9703
SBF [39]	28.7450	0.0625	0.1264	3.7240	86.8112
SRAD 9]	27.6271	0.0830	1.5778	3.7240	112.2950
Frost [5]	28.2110	0.0404	0.1350	3.7240	98.1702
Noise2Void [20]	24.5687	0.3209	0.000 89	3.1642	227.0969
Convolutional autoencoder [17]	29.6552	0.1095	0.1267	3.7240	70.3977

Table 10.2. Quantitative result on speckled image corrupted with $\sigma = 40$ value.

Method	PSNR	EPI	CNR	ENL	MSE
Lee [42]	21.7776	0.5093	0.2148	3.7240	431.8359
Kuan [21]	28.4883	0.0458	0.1758	3.7240	92.0976
SBF [39]	28.3175	0.0434	0.1196	3.7240	95.7912
SRAD [8]	27.6570	0.0684	1.5612	3.7240	111.5253
Frost [5]	28.0267	0.0385	0.1288	3.7240	102.4237
Noise2Void [20]	23.5794	0.3432	0.0215	3.1642	285.1924
Convolutional autoencoder [17]	28.4929	0.0515	0.1247	3.7240	91.9986

proposed in the literature. Further investigations are required to enhance the performance, robustness, and efficiency of these techniques and still, it is a significant research topic. Recently, DL approaches have gained popularity in speckle noise reduction. In this chapter, we provided an overview of significant image despecking methods developed in the past. Also, we discussed the process of designing deep learning methods for image despeckling. Furthermore, the chapter includes illustrative examples of despeckling filters implemented on Python scripts. It is worth mentioning that we have included a few sample experiments, the result evaluation process using both qualitative and quantitative methods are described solely for explanatory purposes.

Acknowledgments

The authors would like to express their sincere gratitude to the authors and creators of the online sources that greatly contributed to the development of this book chapter. Also, the authors would like to acknowledge the scholarly articles, research papers, and websites that provided invaluable insights and information.

References

[1] Ahmed N, Natarajan T and Rao K R 1974 Discrete cosine transform *IEEE Trans. Comput.* **100** 90–3

[2] Ando K, Nagaoka R and Hasegawa. H 2020 Speckle reduction of medical ultrasound images using deep learning with fully convolutional network *Jpn. J. Appl. Phys.* **59**.SK SKKE06

[3] Anoop B N *et al* 2019 Despeckling algorithms for optical coherence tomography images: a review *Advanced Classification Techniques for Healthcare Analysis* (Hershey, PA: IGI Global) 286–310

[4] Al-Asad J F, Reza A M and Techavipoo U 2014 An ultrasound image despeckling approach based on principle component analysis *Int. J. Image Process. (IJIP)* **8** 156–77

[5] Banerjee S *et al* 2020 A comprehensive survey on frost filter and its proposed variants *2020 5th Int. Conf. on Communication and Electronics Systems (ICCES)* (Piscataway, NJ: IEEE) 109–14

[6] Bioucas-Dias J× M and Figueiredo M× A T 2010 Multiplicative noise removal using variable splitting and constrained optimization *IEEE Trans. Image Process.* **19** 1720–30

[7] Chen Y *et al* 2003 Aggressive region growing for speckle reduction in ultrasound images *Pattern Recognit. Lett.* **24** 677–91

[8] Choi H and Jeong J 2018 Speckle noise reduction in ultrasound images using SRAD and guided filter *2018 Int. Workshop Adv. Image Technol. (IWAIT)* (Piscataway, NJ: IEEE) 1–4

[9] Coup×e P *et al* 2009 Nonlocal means-based speckle filtering for ultrasound images *IEEE Trans. Image Process.* **18** 2221–9

[10] Cui J *et al* 2019 PET image denoising using unsupervised deep learning *Eur. J. Nucl. Med. Mol. Imaging* **46** 2780–9

[11] Dong Z *et al* 2020 Optical coherence tomography image denoising using a generative adversarial network with speckle modulation *J. Biophotonics* **13** e201960135

[12] Farhadiani R, Homayouni S and Safari A 2019 Hybrid SAR speckle reduction using complex wavelet shrinkage and non-local PCA-based filtering *IEEE J. Sel. Top. Appl. Earth Obs. Remote Sens.* **12** p 1489–96

[13] de Fatima Carvalho Ferreira A and Fernandes D 2000 Speckle filter for Weibull-distributed SAR images *IGARSS 2000. IEEE 2000 Int. Geoscience and Remote Sensing Symp.. Taking the Pulse of the Planet: The Role of Remote Sensing in Managing the Environment. Proc. (Cat. No. 00CH37120)* **vol 2** (Piscataway, NJ: IEEE) 642–4

[14] Gai S *et al* 2018 Speckle noise reduction in medical ultrasound image using monogenic wavelet and Laplace mixture distribution *Digital Signal Process.* **72** 192–207

[15] Göreke V 2023 A novel method based on wiener filter for denoising poisson noise from medical x-ray images *Biomed. Signal Process. Control* **79** 104031

[16] Gravel P, Beaudoin G and De Guise J A 2004 A method for modeling noise in medical images *IEEE Trans. Med. Imaging* **23** 1221–32

[17] Karaŏglu O, Şakir Bilge H and İhsan Uluer 2022 Removal of speckle noises from ultrasound images using five different deep learning networks *Eng. Sci. Technol. Int. J.* **29** 101030

[18] Kittisuwan P 2018 Speckle noise reduction of medical imaging via logistic density in redundant wavelet domain *Int. J. Artif. Intell. Tools* **27** 1850006

[19] Koundal D, Gupta S and Singh S 2016 Nakagami-based total variation method for speckle reduction in thyroid ultrasound images *Proc. Inst. Mech. Eng. H: J. Eng. Med.* **230** 97–110

[20] Krull A, Buchholz T-O and Jug F 2019 Noise2void-learning denoising from single noisy images *Proc. of the IEEE/CVF Conf. on Computer Vision and Pattern Recognition* (Piscataway, NJ: IEEE) 2129–37

[21] Lee J-S *et al* 1994 Speckle filtering of synthetic aperture radar images: a review *Remote Sens. Rev.* **8** 313–40

[22] Liu S *et al* 2019 Deep learning in medical ultrasound analysis: a review *Engineering* **52** 261–75

[23] Loupas T, McDicken W N and Allan P L 1989 An adaptive weighted median filter for speckle suppression in medical ultrasonic images *IEEE Trans. Circuits Syst.* **36** 129–35

[24] Maier A and Fahrig R 2015 GPU denoising for computed tomography *Graphics Processing Unit-Based High Performance Computing in Radiation Therapy* **1** 113–28

[25] Maity A *et al* 2015 A comparative study on approaches to speckle noise reduction in images *2015 Int. Conf. on Computational Intelligence and Networks* (Piscataway, NJ: IEEE) 148–55

[26] Manj×on J V and Coupe P 2018 MRI denoising using deep learning *PatchBased Techniques in Medical Imaging: 4th Int. Workshop, Patch-MI 2018, Held in Conjunction with MICCAI 2018, Granada, Spain, September 20, 2018, Proc. 4* (Berlin: Springer) 12–9

[27] Ning C-Y, Liu S-f and Qu M 2009 Research on removing noise in medical image based on median filter method *2009 IEEE Int. Symp. IT Med. Edu.* **vol 1** (Piscataway, NJ: IEEE) 384–8

[28] Ouahabi A 2013 A review of wavelet denoising in medical imaging *2013 8th Int. Workshop on Systems, Signal Processing and their Applications (WoSSPA)* (Piscataway, NJ: IEEE) 19–26

[29] Seabra J C R 2011 Medical ultrasound B-mode modeling, de-speckling and tissue characterization assessing the atherosclerotic disease *PhD Thesis* (Instituto Superior T×ecnico)

[30] Shi F *et al* 2019 DeSpecNet: a CNN-based method for speckle reduction in retinal optical coherence tomography images *Phys. Med. Biol.* **64** 175010

[31] Simard M *et al* 1998 Analysis of speckle noise contribution on wavelet decomposition of SAR images *IEEE Trans. Geosci. Remote Sens.* **36** 1953–62

[32] Singh M, Singh S and Kansal S 2009 Comparative Analysis of spatial filters for speckle reduction in ultrasound images *2009 WRI World Congress on Computer Science and Information Engineering* **vol 6** (Piscataway, NJ: IEEE) 228–32

[33] Slabaugh G *et al* 2006 Ultrasound-specific segmentation via decorrelation and statistical region-based active contours *2006 IEEE Computer Society Conf. on Computer Vision and Pattern Recognition (CVPR'06)* vol 1 (Piscataway, NJ: IEEE) 45–53

[34] Smitha A and Jidesh P 2021 A nonlocal deep image prior model to restore optical coherence tomographic images from gamma distributed speckle noise *J. Mod. Opt.* **68** 1002–17

[35] Sudeep P V *et al* 2016 Speckle reduction in medical ultrasound images using an unbiased non-local means method *Biomed. Signal Process. Control* **28** 1–8

[36] Swamy S and Kulkarni P K 2020 A basic overview on image denoising techniques *Int. Res. J. Eng. Technol.* **7** 850–7

[37] Tajmirriahi M *et al* 2021 A lightweight mimic convolutional auto-encoder for denoising retinal optical coherence tomography images *IEEE Trans. Instrum. Meas.* **70** 1–8

[38] Tao Z, Tagare H D and Beaty J D 2006 Evaluation of four probability distribution models for speckle in clinical cardiac ultrasound images *IEEE Trans. Med. Imaging* **25** 1483–91

[39] Tay P C, Acton S T and Hossack J A 2006 Ultrasound despeckling using an adaptive window stochastic approach *2006 Int. Conf. on Image Processing* (Piscataway, NJ: IEEE) 2549–52

[40] Yu Z, Rahman M A and Jha A K 2022 Investigating the limited performance of a deep-learning-based SPECT denoising approach: an observerstudy-based characterization *Medical Imaging 2022: Image Perception, Observer Performance, and Technology Assessment* (Bellingham, WA: SPIE) vol 12035

[41] Zhou Y *et al* 2021 Speckle noise reduction for OCT images based on image style transfer and conditional GAN *IEEE J. Biomed. Health Inform.* **26** 139–50

[42] Zhu J, Wen J and Zhang Y 2013 A new algorithm for SAR image despeckling using an enhanced Lee filter and median filter *2013 6th Int. Congress Image Signal Process. (CISP)* vol 1 (Piscataway, NJ: IEEE) 224–8

IOP Publishing

Image Processing with Python
A practical approach
Irshad Ahmad Ansari and Varun Bajaj

Chapter 11

Application of image processing and machine learning techniques for vegetation cover classification in precision agriculture

Atiya Khan, C H Patil, Amol D Vibhute and Shankar Mali

An overview of the use of image processing and machine learning approaches for classifying vegetation cover in precision agriculture is given in this chapter. By using data-driven strategies, precision agriculture seeks to increase crop output and lower input costs. Vegetation cover classification using machine learning algorithms and image processing methods can provide important information about the development and health of crops. Classifying the vegetation cover is essential for evaluating crop health, locating weed infestations, and tracking the dynamics of the entire vegetation.

We used multi-temporal Sentinel-2 imagery for areas of interest in the Nagpur district. Visual and numerical comparisons between extracted fields and ground reference data were made. The method demonstrated dependability in producing polygons that nearly matched reference data. The NDVI vegetation index offers data on the significance or dominance of the vegetation cover and enables modelling of the various crops' phenological stages. NDVI calculation of ROI areas highlighted types of vegation in the study area.

This chapter addresses the various data collection, preprocessing, feature extraction, model training, and prediction stages involved in precision agriculture utilising image processing and machine learning. This study covers the literature and studies that have already been done on the use of image processing and machine learning methods for classifying vegetation in precision agriculture. It talks about the benefits, difficulties, and probable directions this field could go in the future. The results show how effective these methods could be in advancing sustainable farming and enhancing agricultural practises. The chapter concludes with an examination of the challenges and potential outcomes of using algorithms for image processing and machine learning in precision agriculture.

11.1 Introduction

An essential component of precision agriculture is the classification of vegetation cover. It entails identifying and categorising various forms of vegetation cover in fields of agriculture using algorithms and remote sensing techniques. The objective is to precisely map and track the temporal and spatial variations in the amount of vegetation, which can aid farmers in streamlining crop management techniques, raising yields, and lowering input costs.

In order to classify the vegetation cover of an agricultural area, high-resolution images satellite or aerial photography is often acquired. Then, multiple algorithms and methods based on machine learning are used to process this images in order to identify and categorise various types of vegetation cover, including crops, weeds, and bare soil. Following that, decisions on crop management practises, such as irrigation, fertilisation, and pest control, can be made using the generated vegetation cover maps.

Classifying vegetation cover for precision agriculture has several advantages, one of which is its capacity to give farmers fast and precise information regarding crop health and development. With the aid of this information, it is possible to locate fields that need to be improved, such as those with weed infestations, water stress, or low soil fertility. Farmers may be proactive in improving crop health and output by treating these problems before they become serious.

Overall, classifying vegetation cover is a useful tool for precision farming. Farmers may acquire insightful information on the health and development of their crops and make data-driven decisions that maximise productivity and profitability by utilising the most recent developments in the field of remote sensing and machine learning.

To fulfil future food demand and accomplish the sustainable development goals (SDGs), agricultural production must improve while its environmental effect must be decreased. Precision farming includes a range of technological characteristics, such as machine learning, satellite data, drone applications, navigation, and communication, with an emphasis on helping farmers and a healthy environment accomplish sustainability, climate-related goals, and profitability [1]. Technology has the capacity to learn from experiences thanks to machine learning. We can extract the most significant findings from the vast amounts of crop field data using statistical techniques and machine learning techniques. It makes hidden patterns and connections between factors impacting horticulture, such as temperature, soil salinity, and humidity, apparent. SVM regression, logistic regression and artificial neural networks are the most popular and associated machine learning approaches for predicting agricultural diseases and pests when weather data is studied [2]. Agricultural insurance and agricultural yield modelling were the main consumption sectors, with crop type mapping functioning as the primary emphasis. The need for stakeholders to have access to geographic data regarding crop kind, crop health, and crop stress at the field level has increased with the focus on crop insurance in India. Large regions (districts or states) must also have access to this information, and it must contain details on all the crops farmed there [3]. For the management of agriculture, timely and precise classification of crops is of utmost importance. When using Sentinel-2 data, users can select the optimal spectral band pairings and temporal frame for crop mapping by comprehending how spectral-temporal data affect crop classification. In this study [4], the random forest algorithm was used to create a crop classification map with a 10 m spatial resolution using multi-temporal Sentinel-2 data that were collected during the growing season in 2019 [4]. The majority of cropland mapping systems currently in use are pixel-based time series analyses of high precision remote sensing data. This work assesses the performance of the time-weighted dynamic time warping (TWDTW) approach on the object-based and pixel-based classifications of distinct crop types in three separate study areas [5]. Since so many elements, such as crop genetics, environmental factors, management practises, and their interconnections, affect crop output, forecasting it is quite challenging. A deep learning framework using recurrent neural networks and convolutional neural networks is provided in study work in [6] in order to estimate crop yield based on environmental data and management practises. In-depth research [7] on convolutional neural networks (CNN) for hyperspectral image classification (HSIC) has shown that they are efficient at utilising combined spatial and spectral information, but suffer from slower generalisation performance and learning speed as a result of hard labels and non-uniform label distribution. The aforementioned problems have been resolved using a variety of regularisation techniques [7]. The process of classification has been made easier by hyperspectral imaging (HSI), which measures the reflectance near-infrared (NIR), over visible (VIS), and shortwave infrared (SWIR) [8] wavelengths. Hyperspectral imaging has a range of applications, including agriculture, even at a low level. In each region, bands were quantized and chosen using measures including entropy, modified normalised difference water index (MNDWI) and normalised difference vegetation index (NDVI). To map the chosen crops, a convolutional neural network was created using the finer generated sub-cube [9]. In the experiment [10], a two-phase classification was created to demonstrate how well the image classification performed. For the investigation of image classification, the study specifically used a multiple class classification by support vector machine and convolutional neural networks. A supervised learning model called SVM examines the information utilised in classification [10].

Wireless sensors and monitoring systems powered by artificial intelligence (AI) are in high demand and offer accurate information extraction and analysis. The primary objective of this research is to identify the ideal plant development modeleters. This essay promotes intelligent farming and the idea of reducing agricultural risks. Agriculture has always improved, but the wireless AI sensor will set a higher standard for intelligent agriculture [11].

Accurately identifying and classifying various forms of vegetation cover in agricultural areas is the goal of applying image processing and machine learning approaches for vegetation cover classification in precision agriculture. This can be done by utilising remote sensing technology that can produce high-resolution photos of crop fields, like satellite or drone imagery.

Farmers and agronomists can benefit greatly from a classification of vegetation cover by using it to pinpoint fields that need more pest management, fertilisation, or irrigation. Then, with this knowledge, crop yields may be maximised, input costs can be decreased, and sustainability as a whole can be improved.

By automating the process of analysing massive amounts of data, the use of image processing and machine learning techniques can increase the precision and effectiveness of vegetation cover classification. Large datasets of labelled images can be used to train machine learning algorithms to reliably categorise various forms of vegetation cover. Then, additional photos can be processed using these algorithms to deliver real-time data regarding crop growth and health.

Enhancing crop management strategies, raising agricultural output, and reducing environmental effect are the ultimate goals of integrating processing images and machine learning techniques for the classification of vegetation cover in precision agriculture.

11.2 Revolutionizing crop classification in agriculture through artificial intelligence

The agricultural industry is rapidly evolving, with technology playing an increasingly important role in improving efficiency and productivity. One such technology that has the potential to revolutionize agriculture is AI. AI algorithms have already been applied in various fields, including crop classification. By analyzing data from sensors, drones, and satellite imagery, AI can accurately identify and classify different crops in a field. This technology has the potential to enhance the accuracy and speed of crop classification, enabling farmers to make informed decisions and optimize their yield. This section will explore deeper into the impact of AI on crop classification in agriculture and the potential benefits it can bring to farmers and the industry as a whole.

AI technology gives machines computational intelligence so they can learn, comprehend, and respond appropriately to situations. There are countless uses for this topic in different areas of human life. In the fields of healthcare, agriculture, robotics, e-commerce, finance, and automation, intelligent AI programmes are actively being researched. IoT, another growing technology, connects intelligent sensors and gadgets to one another through the internet. The manufacturing industry, solar power plants, agricultural areas, disaster-prone locales, and other disciplines can all be used by these intelligent sensors to gather data for efficient resource usage [12]. To boost both the quantity and the quality of production from crop fields to meet the rising food demand, machine learning will be applied utilising IoT data analytics in the agriculture industry. Such revolutionary developments are upending traditional farming methods and creating the best opportunities while also having a lot of drawbacks [2]. The deep learning technology for predicting agricultural production generally performs and is more accurate than traditional machine learning methods. Based on the variables/parameters utilised in the model, all deep learning algorithms are equally capable of predicting crop yield. However, CNN and LSTM-based deep learning methods are the most successful for predicting agricultural productivity. CNN has the capacity to identify significant factors that may affect the forecast of crop output [13].

11.3 Data collection and preprocessing

11.3.1 Study area

The Nagpur district is situated in Maharashtra state, India, between latitudes of 21.146 633 N and longitudes of 79.088 860 E. The area of the district is around 9897 square kilometres. The town, which is situated between 274.5 and 652.70 metres above sea level, has a 28 percent forest cover. In the Nagpur district, all arable land is divided into three types: rice fields, dry crop fields, and watered or garden fields. Dry agricultural fields are further divided into rabi (late monsoon) and kharif (early monsoon) because of their reliance on the monsoon.

The district's main kharif crops are cotton, jowar, paddy, groundnut, fur, udid, mug, til, castor seed, kulthi ambadi chillies, chavali, cucurbits, brinjals, bhendi, and green vegetables. The middle of June through the middle of July are the kharif crop sowing seasons. Between October and December is usually when these crops are harvested. Beginning in the middle of October and ending in the middle of February is the rabi season. In the centre of the district, where the south-west monsoons are infrequent and unpredictable, rabi crops are particularly crucial. This area comprises the southern half of Ramtek tahsil as well as the eastern portions of Nagpur and Umrer tahsils.

11.3.2 Dataset collection

The visible, mid-infrared, NIR, and thermal infrared regions of the spectrum are covered by the tens to hundreds of extremely small, continuous spectral bands that hyperspectral imaging sensors [14] can gather image data in. For a range of remote sensing applications, these technologies present fresh possibilities for improved differentiation and estimate of biophysical properties. For diverse remote sensing applications, a number of commercial airborne hyperspectral sensors, including CASI, HYDICE, AVIRIS, HyMap, HySpex, and ASIA have been developed [15]. Panchromatic remote sensing cannot gather the spectrum properties and their differences with the same thoroughness and care as hyperspectral remote sensing. This work uses hyperspectral techniques to precisely categorize crops and advance the development of focused hyperspectral uses for agriculture remote sensing, for example monitoring agricultural development and enhancing the administration of the farming industry [16].

The simultaneous achievement of high resolution and wide width on a single satellite results in an imaging width of more than 60 km at a resolution of 2 m. As a result, the GF-1 satellite can offer more detailed visual coverage of items on the ground. Numerous sectors, including ecological environment monitoring, geographical national condition survey and disaster monitoring and analysis, can benefit from the use of the GF-1 satellite image [17]. Due to recent technology improvements, UAVs may now use a number of image collection methods to capture high-quality photographs in many band ranges, including camera sensors that are RGB, multi-spectral, and hyperspectral. As a result, a number of camera sensor parameters, including price and spectral/spatial resolution, should be taken into account depending on the applications. These sensors have demonstrated considerable promise for numerous applications in smart farming and precision agriculture [18]. The Copernicus Sentinels 1 and 2 and NASA's Landsat satellite have routinely gathered photos over the past few years that are synchronised in terms of temporal, spectral and spatial dimensions. The regular and thorough inspection of agricultural activities has been made possible, among other things, by the free and open sharing of such an imaging library. A more accurate modelling of the complex agricultural ecosystems and the generation of knowledge that will promote smart farming, the execution of the common agriculture policy (CAP), and insurance for agriculture have also been made possible by recent developments in AI algorithms and models [19].

The high-dimensional and complex nature of hyperspectral images presents unique challenges and opportunities for developing new image processing and machine learning techniques.

All of the data collections used in the study work can have their full quality photographs viewed and compared using Copernicus satellite data. To view the generated data in the browser, navigate to the area of interest, choose the desired time range, and then check the cloud coverage. Download high resolution photos, experiment with other visualisations, or develop your own. Follow the given steps to download the Sentinel-2 A images through online platform.

1. Open the browser and enter https://scihub.copernicus.eu/ link then go to website Copernicus Open Access Hub and choose Open Hub [32]
2. You need select 'Open Hub' first. The previous page will then appear. In order to use the website, logging in is required. Click the symbol in the page's upper right corner to get started. Registration in advance is necessary. Click 'Sign up' to sign up. The registration form that appears next after clicking 'Sign up' is displayed. Press 'Login' after providing your login information (username) and password (password) (figuress 11.1 and 11.2).

3. To set search criteria, click the symbol shown by the arrow in figure 11.3. It will then display the list as seen in the screenshot below.
   - Choose the following to order the images from your search:
   - Sort the list by Date of ingestion, sensing window, and tile ID.
   - Order by ascending or descending.
   - Sensing period/Ingestion period: The user's choice of dates for their search.

4. These are general search criteria that are used to choose the satellite doing the imaging for each data satellite platform. These satellites share the same characteristics. Leave the field empty to search for photos from both satellites and to broaden the search area: • S2A; • S2B.
5. Product Type—product type, choosing the level of picture processing:
   - S2MSI1C—source; not corrected for atmospheric effects.
6. As shown in figure 11.4, you can toggle the magnifier. Within the chosen polygon, it will look for any photographs where the search criteria are true. The photographs that were discovered will be listed. Choose the fourth picture on

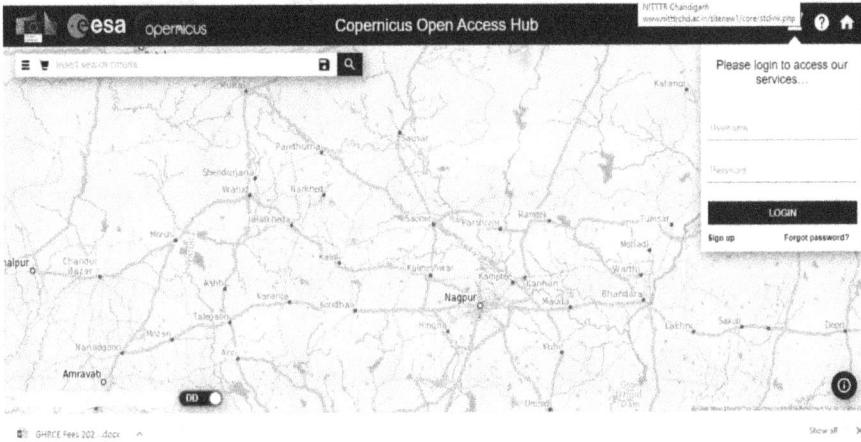

Figure 11.1. Select the area and login. Contains modified Copernicus Sentinel data [2024] processed by Sentinel Hub/IOP Publishing.

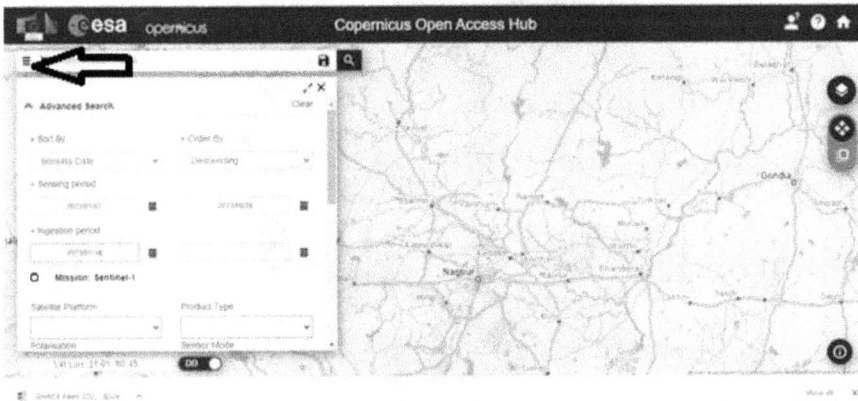

Figure 11.2. Option selection for searching and downloading images for ROI. Contains modified Copernicus Sentinel data [2024] processed by Sentinel Hub/IOP Publishing.

the list: Click 'Download Product'. The window will now show up. Choose the download path. Due to the size of the file, this could take a while.

11.3.3 Localization and selection of area of interest (AOI) and extraction of NDVI time-series data

To create a time-series of NDVI layers from a Sentinel-2 tile with a range of land cover classes, such as water, urban area, cropped area, and various plant classes, to test the aforementioned method. Low NDVI values indicate a paucity of green vegetation (such as a field right after planting), whereas high values indicate a dense, green canopy. NDVI is a measure of vegetative growth. Using the website Scihub Copernicus, we will first choose a number of scenes with a cloud cover of no more

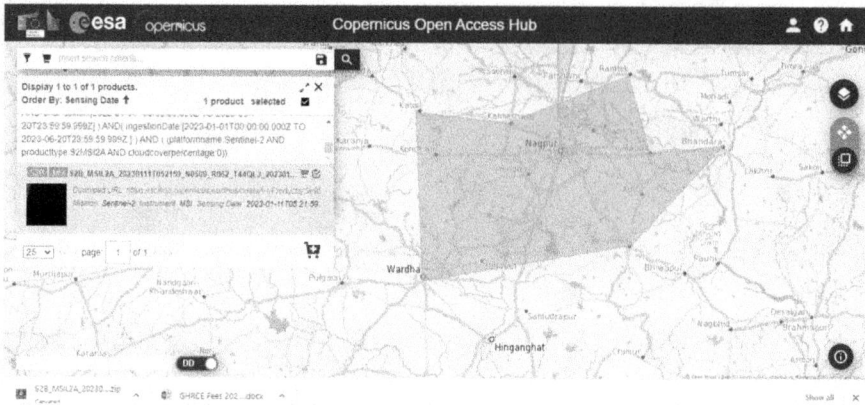

Figure 11.3. Sentinal 2 image product download option. Contains modified Copernicus Sentinel data [2024] processed by Sentinel Hub/IOP Publishing.

Figure 11.4. Band 4 as an output is displayed.

than 15%. We will manually create polygons corresponding to these general land cover categories using QGIS and rasterize them to enable the creation of land cover masks and the extraction of pixel-wise NDVI time-series for any given area in order to guarantee a representative sampling of land cover classes.

11.4 Feature extraction and selection

Sentinel 2 is a multispectral satellite that the European Spatial Agency (ESA) launched. It has 13 bands and a spatial resolution of 10–60 metres. Bands 2, 3, 4, and 8 in the near infrared spectrum have a spatial resolution of 10 m, and every five days, the entire planet is covered. Due to their high resolution, spatial dispersion, and accessibility, Sentinel 2 images are an invaluable and absolutely free source for agricultural monitoring, land cover classification, and water quality. There are other ways to download Sentinel 2 photos, however the Copernicus Open Access Hub is explored, previsualizing, and images are download from the website https://scihub.copernicus.eu/dhus/#/home.

The key benefit of deep learning methods over conventional features extraction techniques, which necessitate an important level of engineering skill and domain knowledge, is that deep learning features are learned automatically from data using a general-purpose method of learning via deep architecture neural networks. Contrarily, deep learning models with several processing layers are able to learn more effective representations for features of data with multiple degrees of abstraction than the aforementioned unsupervised learning of features techniques, which are often poorly structured models [1].

Convolutional neural networks (CNNs) are deep learning network architectures that get their information directly from data. CNNs are especially helpful for object recognition in images by identifying patterns. They can also be very useful in the classification of non-image data, like time series, signal, and audio data. In addition to reducing the number of parameters, local connectivity and weight sharing help simplify the model, which makes CNNs [20] better suited to processing large numbers of images. CNNs can be one-dimensional, two-dimensional, or three-dimensional by employing convolution kernels with different dimensionalities. Sequence data is fed into 1D-CNNs in order to acquire knowledge and represent sequence connections. Using patch-based 2D-CNNs, spectral and spatial features in images may be learned and represented. Cube-based 3D-CNNs are used to match the spatial, spectral, and temporal features in multi-temporal images [21].

Investigating the viability of representing temporal features with CNNs in multi-temporal classification. A RNN or one of its derivatives, such as the LSTM, is frequently thought of as a reasonable place to start when processing time series because RNNs were first developed to analyse sequential data. In contrast, although theoretically 1D convolution is effective and computationally efficient to recognise temporal patterns at various scales, CNNs [22] were frequently used in the spatial and spectral domains but scarcely across the temporal dimension in remote sensing studies [23].

Python's Rasterio package enables reading, inspecting, visualising, and writing geographic raster data. Satellite images, digital elevation models, and drone photography data products can all be used with the library, which supports GeoTIFF and other spatial raster formats. In an interactive Python environment like a Jupyter notebook, one or more geographic rasters can be imported using Rasterio. The geographic raster's 'duality' can be maintained by the library, which means that it can manage the gridded components' matrix values in addition to their position and resolution characteristics [24]. The Python function generates true colour and false colour geographic raster pictures in TIFF format for each band composite and visible band (figure 11.5).

```
import rasterio
from rasterio import plot
import matplotlib.pyplot as plt
def plot_rgb_composite(red_band_path, green_band_path,
blue_band_path):
```

```
# Open the bands using rasterio
red_band = rasterio.open(red_band_path, driver=
'JP2OpenJPEG')
green_band          =          rasterio.open(green_band_path,
driver='JP2OpenJPEG')
blue_band           =          rasterio.open(blue_band_path,
driver='JP2OpenJPEG')

# Show multiple band representation (RGB composite)
fig, (ax_red, ax_green, ax_blue) = plt.subplots(1, 3,
figsize=(15, 5))
plot.show(red_band, ax=ax_red, cmap='Reds', title='Red
Band')
plot.show(green_band, ax=ax_green, cmap='Greens',
title='Green Band')
plot.show(blue_band, ax=ax_blue, cmap='Blues',
title='Blue Band')
fig.suptitle('Sentinel-2 RGB Composite', fontsize=16)
plt.show()

if __name__ == '__main__':
# Define the paths to the individual bands of the
Sentinel-2 image
red_band_path     =     'c://Dataset_CropClassification//
T44QLJ_20220118T051141_B02.jp2'
green_band_path   =     'c://Dataset_CropClassification//
T44QLJ_20220118T051141_B03.jp2'
blue_band_path    =     'c://Dataset_CropClassification//
T44QLJ_20220118T051141_B04.jp2'
# Call the function to plot the RGB composite
plot_rgb_composite(red_band_path,     green_band_path,
blue_band_path)
```

Reprinted with permission of Saul Montoya (https://hatarilabs.com/ih-en/sentinel2-images-explotarion-and-processing-with-python-and-rasterio)

True Color RGB (4, 3, 2): True colour composite creates a naturally coloured image that accurately depicts the Earth as people would naturally see it by employing the visible light bands blue (B02), red (B04), and green (B03) in the corresponding red, green, and blue colour channels (figure 11.6).

False Color RGB (8,4,3):- Utilising the red, green, and near-infrared bands together, false colour imaging is demonstrated. Near-infrared, red, and green bands are frequently used to make false-color composites. It is most typically used to assess plant density and health since plants absorb red light while reflecting green and near infrared light. Due to the fact that plants reflect near infrared light more than green

Figure 11.5. Representation of three bands (RED, GREEN, AND BLUE) using Rasterio and Matplotlib.

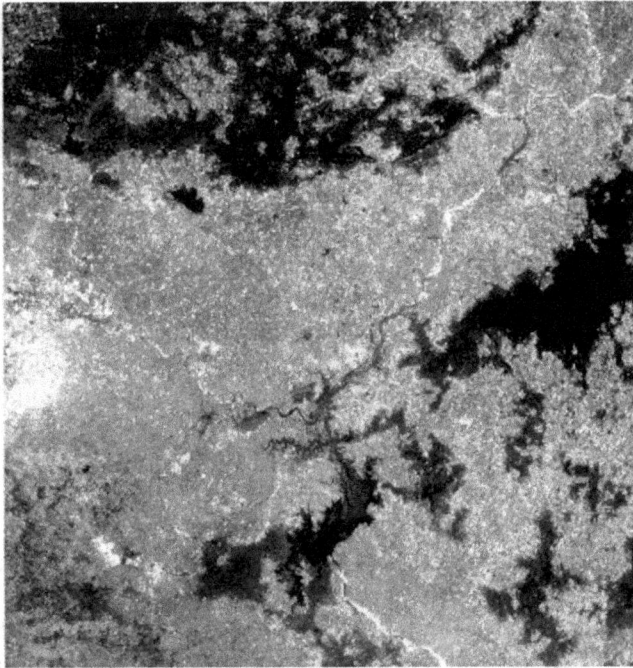

Figure 11.6. True Color representation of Image.

does, land covered with vegetation appears deep red. Darker red is associated with denser plant growth. Cities and the exposed land look to be grey or tan, while the water appears to be blue or black (figure 11.7).

Image feature extraction is the process of using a computer to extract the important data from an image. The feature extraction approach is employed in this phase when it is determined that the features included in the image's pixels are appropriate for categorization [25]. Hyperspectral (HS) data delivers richer spectral information with many narrow bands (i.e., 1–10–nm spectral resolution), in contrast to multispectral data's coarser spectral bands (i.e., 30–100–nm spectral resolution).

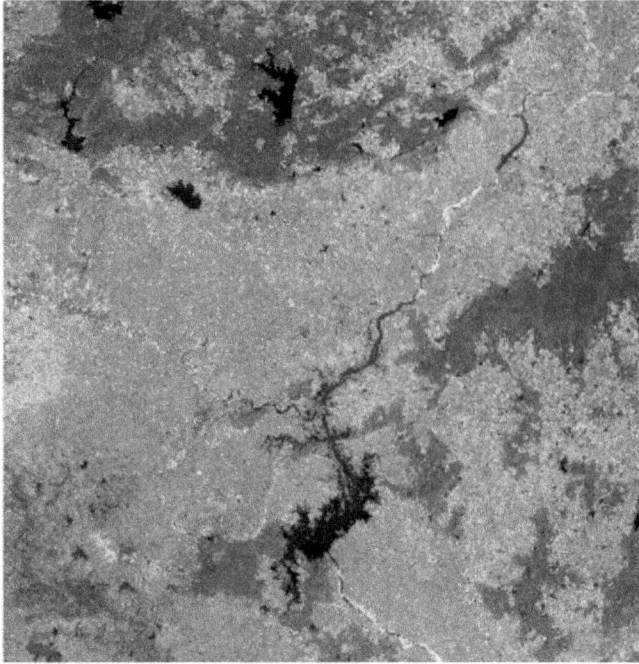

Figure 11.7. False color representation of image.

Numerous narrow spectral bands offer a wealth of spectral information and aid in classifying various land cover and surface features of the Earth (such as the types of plants and crops, soil, and minerals). Assessment of crop or vegetation status, disease detection, estimate of biophysical factors, and monitoring of water quality are further applications that benefit from the fine spectrum information of HS data [26],

Equation (11.1) illustrates how the Normalised Difference Vegetation Index (NDVI) gauges plant health in terms of greenness density. In the field of remote sensing, this vegetation index is frequently utilised. The NDVI has a scale of +1 to −1. The NDVI value of dead plants is 1, whereas that of healthy plants ranges from 0.65 to 1 [9].

The processing of S2 images resulted in products based on the NDVI, a vegetation index now serves as the most commonly used vegetation index in remote sensing, indicating the significance and dominance of vegetation on captured images [27].

Each band in the NIR area is quantified based on NDVI, and the top m2(d2) bands from Partition#2 are chosen based on a threshold limit value of δ2.

$$\text{NDVI} = \frac{(\text{NIR} - \text{RED})}{(\text{NIR} + \text{RED})} \tag{11.1}$$

To calculate the NDVI from a Sentinel-2 image follow the steps given below:

(i) The Numpy library is used for numerical computations, and the Rasterio library is used to read and manipulate raster data.
(ii) Specify the path to the Sentinel-2 image in TIFF format.

(iii) Using the `rasterio.open()` function, the Sentinel-2 image is opened as a Rasterio dataset (`src`). The red band is read from the image using `src.read(3)` where 3 represents the band index. Similarly, the near-infrared band is read using `src.read(8)`.

(iv) The red band values are subtracted from the near-infrared band values, and their sum is divided to determine the NDVI. A new array called ndvi is created as a result of this algorithm, which calculates the NDVI value for each matching pixel in the bands.

(v) The NDVI array is saved as a new GeoTIFF image using `rasterio.open()` with the `'w'` mode (write mode). The output path is specified as `'path_to_save_ndvi_image.tif'`. In the `open()` function, the necessary metadata such as width, height, coordinate reference system (crs), transform, and data type are provided from the source image (`src`). The `count=1` argument specifies that there is only one band in the output image. Finally, `dst.write(ndvi, 1)` writes the NDVI array to the output file.

```
import numpy as np
import rasterio
image_path1      =      'c://Dataset_CropClassification//
Sentinal_raster_18_Jan2022.tif'
with rasterio.open(image_path1) as src:
# Read the red and near-infrared bands from the image
red_band1 = src.read(3) # Red band is typically at index 3
nir_band1 = src.read(4) # Near-infrared band is typi-
cally at index 8
ndvi1 = (nir_band1 - red_band1) / (nir_band1 + red_band1)
# Convert the NDVI array to the specified data type
ndvi1  =  np.nan_to_num(ndvi1,  nan=-9999,  posinf=-9999,
neginf=-9999).astype('int32')
# Save NDVI as a new GeoTIFF file
output_path    =    'c://Dataset_CropClassification//NDVI.
tif'
with rasterio.open( output_path, 'w', driver='GTiff',
width=src.width, height=src.height, count=1, crs1=src.
crs, transform1=src.transform, dtype='int32') as dst:
dst.write(ndvi1, 1)
```

Reprinted with permission of Bhavesh Patidar (https://github.com/bhavesh907/Crop-Classification).

The figure 11.8 shows the results after executing the code and displaying the NDVI index for the study area.

11-13

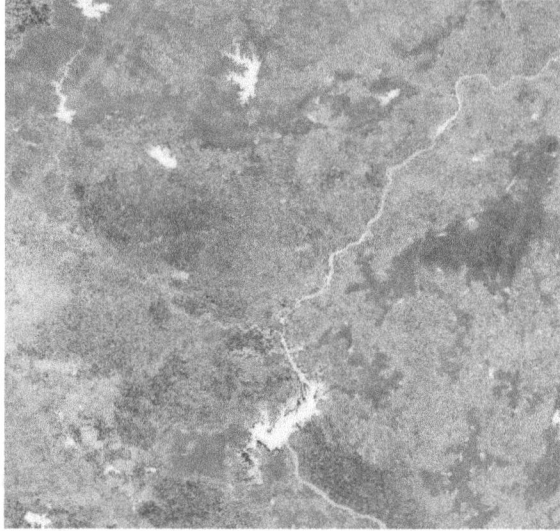

Figure 11.8. NDVI index displaying vegetated section in the study area.

11.5 Methodology

11.5.1 Data loading and preprocessing:

The source code proceeds by importing the required libraries, including Pandas, GDAL, and Numpy. Additionally, it imports the gc module for garbage collection and the operator module for sorting.

By enabling Python exceptions and registering all GDAL drivers, the script configures GDAL. There are defined paths for the dataset and the region of interest (ROI). GDAL is used to open the ROI raster dataset, and the ROI data is then read as a NumPy array. The code finds the distinct classes that are present in the ROI array and uses a dictionary comprehension to calculate the number of pixels for each class. Based on the counts, the classes are arranged in ascending order with respect to their associated pixel counts [28].

   (i) Imports the Python library like csv, Rasterio, and Numpy libraries that are required.

  (ii) Using the image_path variable, it defines the location of an input Sentinel-2 image in TIFF format.

 (iii) The image file supplied by image_path is opened using Rasterio, and the resulting Rasterio dataset is then assigned to the src variable.

 (iv) The image_data variable receives the image data (bands) from the src dataset.

  (v) Using numpy.reshape(), it converts the image_data from a 3D array (rows, columns, and bands) to a 2D array (rows*columns, bands). This procedure is used to get the data ready for exporting as a CSV file.

(vi) Using the output_csv_file variable, it defines the location and name of the output CSV file.

```
CSV    file    saved:    c://Dataset_CropClassification//
path_to_save_output_csv_file.csv
```

(vii) It uses open() to open the output CSV file in write mode and then sets the file object as the value of the csv_file variable. To maintain uniform line endings across platforms, the newline=' parameter is available.

(viii) To write data to the CSV file, it constructs the writer object of type csv.writer.

(ix) It writes the reshaped data (reshaped_data) to the CSV file using the writerows() method of the csv.writer object.

(x) Finally, it outputs a message confirming the CSV file's successful saving along with the filename and path.

```python
import Numpy as np
import rasterio
import csv
def convert_tiff_to_csv(image_path, csv_output_file):
  # Open the image using rasterio
  with rasterio.open(image_path) as src:
  # Read the image data (bands)
  image_data = src.read()
  # Reshape the image data to a 2D array (rows, columns)
for CSV conversion
  rows, cols, bands = image_data.shape
  reshaped_data = np.reshape(image_data, (rows * cols,
bands))
  # Write a CSV file with the reshaped data
  with open(csv_output_file, 'w', newline=')
as csv_file:
  writer = csv.writer(csv_file)
  for row in reshaped_data:
  writer.writerow(row)
  # Print a message indicating successful CSV file
creation
  print(f'Saved the output CSV file: {csv_output_file}')
  if —name— == '—main—':
  # Specify the path to the Sentinel-2 image in TIFF
format
  image_path    =    'c://Dataset_CropClassification//
vegetated_classes_18_Jan.tif'
```

```
  # Name and path of the CSV output file
  csv_output_file = 'c://Dataset_CropClassification//
path_to_save_CSV_output_file.csv'
  # Call the function to convert the TIFF image to CSV
  convert_tiff_to_csv(image_path, csv_output_file)
```

11.5.2 Using deep learning model

The code then imports the necessary libraries, including Numpy, Pandas, train_-
test_split from, LabelEncoder, StandardScaler and sklearn.model_selection from
sklearn.preprocessing, along with a large number of modules from Keras.

```
import numpy as np
import rasterio
import tensorflow as tf
def   main(preprocessed_csv_path,   model_path,   out-
put_tiff_path, image_meta, original_image_shape):
  # Read the preprocessed data from the CSV file
  preprocessed_data = np.loadtxt(preprocessed_csv_path,
delimiter=',')
  # Reshape the preprocessed data back to the original
image shape (rows, columns, bands)
  rows, cols, bands = original_image_shape # Replace with
the actual shape of the original image
  preprocessed_image   =   np.reshape(preprocessed_data,
(rows, cols, bands))
  # Load your trained deep learning model
  model = tf.keras.models.load_model(model_path)
  # Perform prediction on the preprocessed image
  predicted_image   =   model.predict(np.expand_dims(pre-
processed_image, axis=0))
  # Adjust the shape of the predicted image (if neces-
sary)
  predicted_image = np.squeeze(predicted_image, axis=0)
  # Write the predicted image to a TIFF file
  with  rasterio.open(output_tiff_path,  'w',  **image_-
meta) as dst:
  dst.write(predicted_image,  indexes=1)  # Assuming  a
single-band predicted image
  if -name- == '-main-':
  # Specify the path to the preprocessed image (in CSV
format)
  preprocessed_csv_path =
'c://Dataset_CropClassification//
```

```
path_to_save_output_csv_file.csv'
  # Specify the path to the trained deep learning model
  model_path = 'path_to_trained_model.h5'
  # Specify the output path and filename for the pre-
dicted image (in TIFF format)
  output_tiff_path = 'output.tif'
  # Replace 'image_meta' and 'original_image_shape' with
actual metadata and image shape information
  image_meta = {} # Add the necessary metadata
  original_image_shape = (100, 100, 4) # Replace with the
actual shape of the original image
  # Call the main function with the specified parameters
  main(preprocessed_csv_path,      model_path,      out-
put_tiff_path, image_meta, original_image_shape)
  ### Implementing Model 2 for crop Classification
  import numpy as np
  import pandas as pd
  df1    =    pd.read_csv('C:/Dataset_CropClassification/
path_to_save_CSV_output_file.csv')
  labels = df1.iloc[:, -1]
  data1 = df1.drop(df1.columns[50], axis=1)
  encoder1 = LabelEncoder()
  encoder1.fit(labels)
  encoded_Y1 = encoder1.transform(labels)
  dummy_y1  =  tf.keras.utils.to_categorical(encoded_Y1,
num_classes=5)
  # Create test sets and training sets from the data.
  X1_train, X1_test, y1_train,
y1_test = train_test_split(data1,dummy_y1,
test_size=0.05,random_state
=42,shuffle=True)
  # Standardize the input features
  from keras.wrappers.scikit_learn import
KerasClassifier
  from keras.utils import np_utils
  from sklearn.preprocessing import LabelEncoder
  from sklearn.pipeline import Pipeline
  from sklearn.model_selection import train_test_split
  from keras.models import Sequential
  from keras.layers import Dense
  from sklearn.model_selection import cross_val_score
  from sklearn.model_selection import KFold
  from sklearn.preprocessing import StandardScaler
```

```
sc_X1 = StandardScaler()
X1_train = sc_X1.fit_transform(X1_train)
X1_test = sc_X1.transform(X1_test)
import numpy
 import pandas
 from keras import regularizers
 from keras import initializers
 from keras.layers import Dense, Dropout, Activation
 from keras.optimizers import SGD
 seed = 7
 numpy.random.seed(seed)
 model11 = Sequential()
 model11.add(Dense(200, input_shape=(50,),
activation='relu',
 kernel_regularizer=regularizers.l2(1e-5),
 kernel_initializer=initializers.glorot_normal(seed=-
seed),
 bias_initializer='zeros'))
 model11.add(Dropout(0))
 model11.add(Dense(5, activation='softmax'))
 model11.summary()
 Model: "sequential_16"
```

Model: "sequential_16"

Layer (type)	Output Shape	Param #
dense_27 (Dense)	(None, 200)	10200
dropout_13 (Dropout)	(None, 200)	0
dense_28 (Dense)	(None, 5)	1005

Total params: 11,205
Trainable params: 11,205
Non-trainable params: 0

```
sgd = keras.optimizers.Adadelta()
model11.compile(loss='categorical_crossentropy',optimi-
zer=sgd,metrics=['acc'])
 # Preprocess input data
 #X1_train = pad_sequences(X1_train, maxlen=50)
 #X1_test = pad_sequences(X1_test, maxlen=50)
```

```
#y1_test = pad_sequences(y1_test, maxlen=50)
#y1_test = pad_sequences(y1_test, maxlen=50)
# Build the initial model.
model11.fit(X1_train,  y1_train,  epochs=10,  batch_-
size=100, shuffle=True)
score11 = model11.evaluate(X1_test, y1_test, batch_-
size=100)
print('\n%s:  %.2f%%'  %  (model11.metrics_names[1],
score11[1]*100))
```

```
Epoch 1/10
110/110 [==============================] - 1s 3ms/step - loss: 1.6
062 - acc: 1.0000
Epoch 2/10
110/110 [==============================] - 0s 4ms/step - loss: 1.6
059 - acc: 1.0000
Epoch 3/10
110/110 [==============================] - 1s 5ms/step - loss: 1.6
056 - acc: 1.0000
Epoch 4/10
110/110 [==============================] - 1s 6ms/step - loss: 1.6
053 - acc: 1.0000
Epoch 5/10
110/110 [==============================] - 1s 5ms/step - loss: 1.6
049 - acc: 1.0000
Epoch 6/10
110/110 [==============================] - 1s 6ms/step - loss: 1.6
045 - acc: 1.0000
Epoch 7/10
110/110 [==============================] - 1s 6ms/step - loss: 1.6
041 - acc: 1.0000
Epoch 8/10
110/110 [==============================] - 1s 6ms/step - loss: 1.6
037 - acc: 1.0000
Epoch 9/10
110/110 [==============================] - 0s 4ms/step - loss: 1.6
032 - acc: 1.0000
Epoch 10/10
110/110 [==============================] - 0s 3ms/step - loss: 1.6
027 - acc: 1.0000
```

```
n_classes1 = 5
model21 = Sequential()
model21.add(Conv1D(kernel_size=5,      strides=1,      fil-
ters=32, activation='relu', input_shape=(50, 1)))
model21.add(MaxPooling1D(pool_size=2, strides=2))
model21.add(Conv1D(kernel_size=5,      strides=1,      fil-
ters=64, activation='relu'))
```

```
model21.add(MaxPooling1D(pool_size=2, strides=2))
model21.add(Flatten())
model21.add(Dense(1000, activation='relu'))
model21.add(Dense(n_classes1, activation='softmax'))
model21.summary()
model21.compile(loss='categorical_crossentropy', opti-
mizer=tf.keras.optimizers.Adam(lr=0.001), metrics=
['accuracy'])
```

Model: "sequential_19"

Layer (type)	Output Shape	Param #
conv1d_2 (Conv1D)	(None, 46, 32)	192
max_pooling1d_2 (MaxPooling 1D)	(None, 23, 32)	0
conv1d_3 (Conv1D)	(None, 19, 64)	10304
max_pooling1d_3 (MaxPooling 1D)	(None, 9, 64)	0
flatten_1 (Flatten)	(None, 576)	0
dense_32 (Dense)	(None, 1000)	577000
dense_33 (Dense)	(None, 5)	5005

Total params: 592,501
Trainable params: 592,501
Non-trainable params: 0

```
X1_train_reshaped = X1_train.reshape(X1_train.shape[0],
X1_train.shape[1],1)
X1_test_reshaped=X1_test.reshape(X1_test.shape[0],
X1_test.shape[1],1)
model21.fit(X1_train_reshaped, y1_train, epochs=10,
batch_size=100)
scores21 = model21.evaluate(X1_test_reshaped,y1_test)
print('\n%s: %.2f%%' % (model21.metrics_names[1],
scores21[1]*100))
```

```
Epoch 1/10
110/110 [=============================] - 9s 10ms/step - loss: 1.5243
- accuracy: 1.0000
Epoch 2/10
110/110 [=============================] - 1s 9ms/step - loss: 1.3598 -
accuracy: 1.0000
Epoch 3/10
110/110 [=============================] - 1s 9ms/step - loss: 1.2083 -
accuracy: 1.0000
Epoch 4/10
110/110 [=============================] - 1s 8ms/step - loss: 1.0703 -
accuracy: 1.0000
Epoch 5/10
110/110 [=============================] - 1s 8ms/step - loss: 0.9459 -
accuracy: 1.0000
Epoch 6/10
110/110 [=============================] - 1s 9ms/step - loss: 0.8349 -
accuracy: 1.0000
Epoch 7/10
110/110 [=============================] - 1s 8ms/step - loss: 0.7367 -
accuracy: 1.0000
Epoch 8/10
110/110 [=============================] - 1s 9ms/step - loss: 0.6504 -
accuracy: 1.0000
Epoch 9/10
110/110 [=============================] - 1s 9ms/step - loss: 0.5750 -
accuracy: 1.0000
Epoch 10/10
110/110 [=============================] - 1s 9ms/step - loss: 0.5093 -
accuracy: 1.0000
19/19 [=============================] - 0s 5ms/step - loss: 0.4790 - a
ccuracy: 1.0000

accuracy: 98.00%
```

The above code perform the following tasks:
  (i) Imports the necessary libraries: Pandas, LabelEncoder and StandardScaler from sklearn.preprocessing, train_test_split from sklearn.model_selection, and various components from Keras and TensorFlow.
 (ii) Loads a dataset from a CSV file using pd.read_csv() and assigns it to the DataFrame df.

(iii) Separates the class labels (labels) from the input features (data) in the DataFrame.

(iv) Encodes the class labels using LabelEncoder() to convert categorical labels into numerical values.
 (v) It one-hot encodes the encoded labels using keras.utils.to_categorical() to convert them into a binary matrix representation.

(vi) Splits the data into training and test sets using train_test_split().

(vii) Then it is standardizes the input features using StandardScaler() to ensure they have zero mean and unit variance.

(viii) It builds and compiles the first model (model1) using Sequential() from Keras. The model architecture consists of a dense layer with 200 units, ReLU activation, and L2 regularization. It uses the Adadelta optimizer and categorical cross-entropy loss for training.

(ix) Then preprocesses the input data by applying pad_sequences() to ensure all input sequences have a fixed length of 50.

(x) It trains model1 using the training data (X_train and y_train) with 10 epochs and a batch size of 100.

(xi) Evaluates the trained model1 on the test data (X_test and y_test) and computes the accuracy score.

(xii) Then builds and compiles the second model (model2) using Sequential() from Keras. The model architecture consists of a series of convolutional and pooling layers followed by dense layers. It uses the Adam optimizer and categorical cross-entropy loss.

(xiii) It reshapes the input data (X_train and X_test) to fit the expected input shape of model2.

(xiv) It trains model2 using the reshaped training data (X_train_reshaped and y_train) with 10 epochs and a batch size of 100.

(xv) It evaluates the trained model2 on the reshaped test data (X_test_reshaped and y_test) and computes the accuracy score.

(xvi) Overall, this code builds and trains two different models (model1 and model2) on the given dataset, and evaluates their performance using the accuracy metric. The first model is a fully connected neural network, while the second model is a convolutional neural network.

(xvii) The second model was created by using Sequential(). The base layer is Conv1D, a 1D convolutional layer with kernel size 5, stride 1, and 32 filters. It utilises the ReLU activation function and receives the form (50, 1) as input (figure 11.9).

11.6 Optimizing agriculture with advance machine learning techniques

The use of cutting-edge machine learning techniques for agriculture has grown more and more significant as a result of the growing demand for agricultural products. Farmers may now make data-driven decisions to increase crop output while using fewer resources thanks to machine learning algorithms' success in optimising agricultural practises.

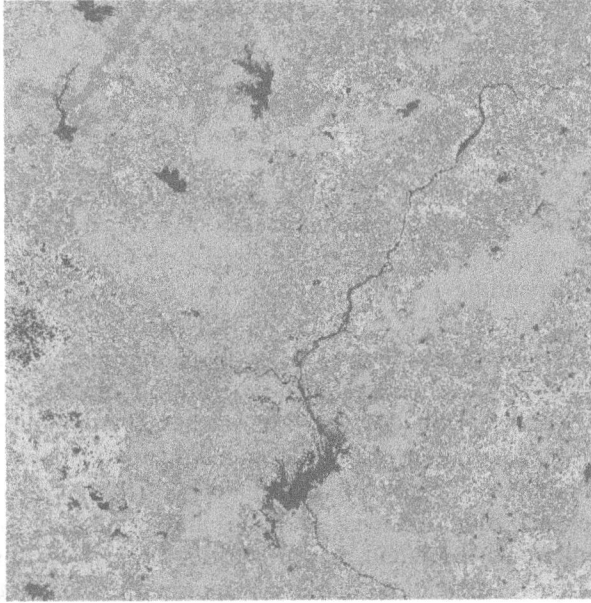

Figure 11.9. Vegetated classes classification.

The quality of the soil, climatic trends, and plant health are just a few of the data that farmers may collect and analyse by utilising machine learning. On crop yields, the best periods for planting and harvesting, irrigation, fertiliser use, pest control, and disease detection, these algorithms can offer precise predictions and suggestions.

Advanced Machine Learning algorithms can also manage huge and complicated datasets, improving the accuracy and precision of predictions and recommendations. Advanced machine learning techniques can be used to optimise agricultural practises in order to make them more effective, profitable, and sustainable while also assuring food security for the expanding global population.

11.7 Applications of machine learning in precision agriculture

11.7.1 Crop yield prediction

Any farmer wants to maximise the profit from their crops and make use of all the advantages that are available. Organizing the crop and figuring out what each field can produce are two ways to do this. Sensing and mapping are the most fascinating developments in this field. This method maps the yield of a field using methods such as imaging and digital image processing.

Machine learning in agriculture is crucially used for crop production prediction. Machine learning models are able to predict potential crop yields by examining past data, weather trends, and other pertinent variables. Farmers are now better equipped to manage crops, allocate resources, and make financial plans as a result of this information. The technique is broken down into the following steps: gathering data, selecting features, model validation and training, forecasting yields, and decision assistance. Although these models have inherent uncertainties, they offer useful tools for taking decisions and can aid farmers in more successfully adapting to changing circumstances.

11.7.2 Soil moisture estimation

One of the most essential components of farming is managing the soil and the water. Data collection helps farmers better understand the water level and soil quality on their fields. Following that, these can be uploaded to a system controlled by computers that, depending on the weather forecast, recommends amounts for applying fertiliser, potential pest management techniques, and watering schedules.

Calculating soil moisture is a significant use of machine learning in agriculture. In agricultural settings, soil moisture levels can be predicted and tracked using a variety of sources of data and machine learning approaches. Planning effective irrigation systems, determining crop water stress, and managing all available water resources all depend on this data. The procedure includes gathering data from sensors or satellite imaging, choosing features, training and validating the model, and estimating the soil moisture content. In agriculture, machine learning models offer useful tools to improve irrigation techniques, preserve water, and increase crop output.

11.7.3 Crop disease detection

Crop management, a broad category of pre-harvest activities, affects future yields. However, this stage of the agricultural lifecycle is one of the most challenging. Increased frequency of droughts, greater temperatures, and unforeseen cycles of soaking and drying, and other conditions may have an effect on crop tolerance. Machine learning technology is often employed to move this step along.For crops to be disease- and weather-resistant, the proper gene sequence is required. Machine learning-based deep learning helps simplify crop breeding. A probabilistic model is created using straightforward algorithms that gather data on plant activity from the field.

A key use in agriculture is the detection of crop diseases using machine learning. Machine learning models can recognise and categorise agricultural diseases using image analysis and pattern recognition techniques, enabling early detection and efficient control. Farmers can then act quickly to control the spread of diseases, reduce crop losses, and make the best use possible of pesticides and fungicides. The method entails collecting plant image data, preparing the data, creating models using various kinds of techniques, like convolutional neural network algorithms (CNNs), verifying the models, and detecting illnesses. Farmers are better able to control illness because to the helpful tools given by machine learning models.

11.7.4 Weed detection and management

Weed control is one of the most crucial jobs for agricultural output. To avoid harm to agricultural output, it is crucial to identify and eradicate the various weed species. Weeding by hand is time-consuming and difficult on the rear end. Farmers used to use herbicides as a last resort, but it has now been proven that they are bad for the environment. Machine learning-programmed robots are a novel option that are currently in development [33].

A prominent use of machine learning in agriculture is weed detection and management. It is feasible to automatically recognise and categorise weeds in agricultural fields by using computer vision techniques and machine learning algorithms. With the aid of this technology, farmers may reduce physical labour, increase crop output, and use herbicides more effectively. Field picture data collecting, data preprocessing, model training with algorithms like CNNs, model validation, and weed detection are all steps in the process. To help farmers control weeds effectively and sustainably, machine learning models offer useful tools. These technologies help farmers make better decisions.

11.7.5 Precision livestock farming

Aspects of machine learning in agriculture that are particularly important are animal welfare and livestock productivity. Numerous fields can benefit from the technology. Assessments of the wellbeing of animals, forecasting animal output, and calculations of the environmental impact of livestock operations are some of these.

Because of this, farmers may better understand the health of their animals by keeping an eye on their vital signs, daily activity levels, and dietary habits. Farmers in Uganda identify infections in animals two days before they start to show symptoms. Users' smartphones or PCs, an RFID reader, and a sensor-equipped chip are all required for this technology to work. The majority of health-related factors, from eating to fertility, can be detected and tracked by the programme in this way.

Machine learning models can help cattle farmers by analysing data from many sources, including sensors, wearable technology, and historical records, to offer insights and enhance decision-making. These models can help with disease outbreak prediction, feed and nutrition optimisation, animal health monitoring, and general livestock productivity and welfare enhancement. Data is gathered from sensors and other devices, preprocessed, then trained into a model using techniques like support vector machines or decision trees. The model is then validated, and management recommendations are made. For livestock farmers looking to improve their management techniques, boost productivity, and protect the welfare of their animals, machine learning models offer useful tools.

11.8 Challenges and opportunities for machine learning in agriculture

Given the high complexity of the data and the scarcity of training samples, HSI presents significant hurdles for supervised classification algorithms. The considerable intraclass variability (as well as interclass similarity) that is frequently present in HSI data, along with these problems, may limit the efficiency of classifiers. Several DL-based architectures have lately been designed, showing considerable promise in HSI data interpretation. To address these shortcomings HSI data offer a number of artefacts in addition to their high dimensionality [29], which makes the classification procedure challenging. Similar to very high-resolution (VHR) images, high intra-class variability in HSI data is caused by uncontrolled variations in the reflectance detected by the spectrometer (typically as a result of shifting atmospheric conditions, cloud-induced occlusions, and changes in clarity, among other environmental interferers) [30].

The computational cost of the model is also impacted by the HSI instruments' addition of duplicate bands. Another difficulty with HSI's spatial resolution is spectral mixing. On the surface of the globe, HSI pixels that have low to average spatial resolution cover a large area, resulting in heterogeneous spectral signatures and strong inter-class similarity in border regions [31].

11.8.1 Limitations and challenges

Despite these restrictions and limits, hyperspectral imaging continues to be an important tool for a variety of applications, including remote sensing, agriculture, environmental monitoring, and medical diagnostics, where the advantages of comprehensive spectral information outweigh the difficulties involved. Hyperspectral imaging's capabilities and accessibility are being improved by ongoing research and technical improvements that overcome some of these constraints.

(I) EO-1 Hyperion images were rare during certain growing seasons.

(II) Data from EO-1 Hyperion have a high signal-to-noise ratio.

(III) High data volume: Hyperspectral images are made up of a lot of small spectral bands, which generates a lot of data. It might be computationally demanding and call for specialised gear and software to manage and process such massive datasets.

(IV) Low spatial resolution: When compared to conventional image sensors, hyperspectral sensors frequently have lesser spatial resolution. This restriction results from the necessity of allocating spectral bands, which lowers pixel resolution. As a result, the specifics of each object or element in the scene could be less clear.

(V) Limited spectral resolution: Although hyperspectral imaging offers comprehensive spectral data, the sensor's spectral resolution is limited. The number of spectral bands can only be as many as the technology that is available, the cost, and the sensor design. The capacity to distinguish between spectral signals that are extremely similar may be hampered by this low spectral resolution.

(VI) Interference caused by the atmosphere, such as light that is absorbed or scattered by it, can have an impact on hyperspectral data. The spectral signatures of items in the scene may be distorted by these atmospheric influences, making correct analysis and interpretation difficult. To lessen these impacts, pre-processing methods like atmospheric correction are frequently used.

(VII) Data storage and transmission: Storing and sending such data can be difficult due to the huge data volume associated with hyperspectral images. Particular attention must be paid to data reduction, storage requirements, and bandwidth needs, particularly in distant sensing applications.

(VIII) Limited temporal coverage: Getting hyperspectral data across a lot of ground can take a while, especially if you're using a moving platform like an aeroplane or satellite. Due to this restriction, several applications that call for frequent temporal coverage cannot access the most recent hyperspectral data.

(IX) Cost: Compared to conventional imaging systems, hyperspectral imaging systems might be more expensive, including sensors and related equipment. The higher cost of technology and data collecting could prevent widespread adoption, especially in places with limited resources.

(X) High computational burden: One of the main issues with DNN is handling a vast amount of data, which necessitates expanding memory bandwidth, high computational cost, and storage utilisation [31].

11.9 Conclusion

The classification process is essential for managing and monitoring changes in land use and cover; remotely sensed imagery could be classified effectively and efficiently using machine learning. Remote sensing has several vital applications, including crop classification and recognition. Machine learning classification approaches have started to appear in recent years for the classification of crops. Accurate and efficient crop classification using remotely sensed data can provide important and crucial information for agricultural research.

For annual crops, the start of the season begins when the rate of increase in NDVI values is greater than the previous successive observations during the period of vegetation growth.

Moreover, we may draw the conclusion that the excellent spectral and spatial resolution of the VNIR and SWIR bands of the Sentinel2 images can provide geologists with a very good opportunity to advance their investigations in the zones with restricted access. The recognition and delineation of various kinds of vegetation in agricultural fields is made possible by the use of segmentation and feature extraction techniques in image processing. Machine learning algorithms use this data as input to automatically classify plant cover based on the attributes they have extracted.

Using deep learning algothims gives better results for land classification for land cover, improves the statistics of land use, and regulates areas that up until now have been difficult to monitor due to their challenging access or the inability to control them using the methods previously used. More modifications are possible so that production can be compared across years, contrasted with environmental factors, and avoided in the case of drought or other unexpected conditions.

In conclusion, the use of machine learning and image processing techniques for identifying vegetation cover in precision agriculture offers considerable promise. It allows farmers to maximise crop management techniques, increase output, and reduce the environmental effect. The agricultural sector as well as society at large will gain from continued development and research in this field as precision agriculture is advanced and widely used.

References

[1] Karfi K E, Fkihi S E, Mansouri L E and Naggar O 2021 Classification of hyperspectral remote sensing images for crop type identification: state of the art *Proc. of the 2nd Int. Conf. on Advanced Technologies for Humanity ICATH* (Setúbal: SciTePress) 11–8

[2] Akhter R and Sofi S A 2022 Precision agriculture using IoT data analytics and machine learning *J. King Saudi Univ.—Comput Inf. Sci.* **34** 5602–18

[3] Gumma M K, Tummala K, Dixit S, Collivignarelli F, Holecz F, Kolli R N and Whitbread A M 2022 Crop type identification and spatial mapping using sentinel-2 satellite data with focus on field-level information *Geocarto. Int.* **37** 1833–49

[4] Yi Z, Jia L and Chen Q 2020 Crop classification using multi-temporal sentinel-2 data in the shiyang river basin of China *Remote. Sens.* **12** 4052

[5] Belgiu M and Csillik O 2018 Remote sensing of environment sentinel-2 cropland mapping using pixel-based and object-based time-weighted dynamic time warping analysis *Remote. Sens. Environ.* **204** 509–23

[6] Khaki S, Wang L and Archontoulis S V 2020 A CNN-RNN framework for crop yield prediction *Front. Plant. Sci.* **10** 1–14

[7] Ahmad M, Mazzara M and Distefano S 2021 Regularized CNN feature hierarchy for hyperspectral image classification *Remote. Sens.* **13** 1–4

[8] Fal S, Maanan M, Baidder L and Rhinane H 2019 The contribution of Sentinel-2 satellite images for geological mapping in the south of Tafilalet basin *Int. Arch. Photogramm. Remote Sens. Spatial Inf. Sci.* **XLII-4/W12** 75–82

[9] Agilandeeswari L, Prabukumar M, Radhesyam V, Phaneendra K L N B and Farhan A 2022 Crop classification for agricultural applications in hyperspectral remote sensing images *Appl. Sci.* **12** 1670

[10] Wan S, Yeh M L and Ma H L 2021 An innovative intelligent system with integrated CNN and SVM: considering various crops through hyperspectral image data *ISPRS Int. J. Geo-Inform.* **10** 242

[11] Murugamani C, Shitharth S, Hemalatha S, Kshirsagar P R, Riyazuddin K, Naveed Q N, Islam S, Mazher Ali S P and Batu A 2022 Machine learning technique for precision agriculture applications in 5G-based internet of things *Wirel. Commun. Mob. Comput.* **2022** 6534238

[12] Sharma A, Jain A, Gupta P and Chowdary V 2021 Machine learning applications for precision agriculture: a comprehensive review *IEEE Access* **9** 4843–73

[13] Muruganantham P, Wibowo S, Grandhi S, Samrat N H and Islam N 2022 A systematic literature review on crop yield prediction with deep learning and remote sensing *Remote Sens.* **14** 1990

[14] Khan A, Vibhute A D, Mali S and Patil C H 2022 A systematic review on hyperspectral imaging technology with a machine and deep learning methodology for agricultural applications *Ecol. Inform.* **69** 101678

[15] Yang C 2015 *Hyperspectral Imaging Technology in Food and Agriculture* ed B Park and R Lu (New York: Springer) 289–304

[16] Wu H, Zhou H, Wang A and Iwahori Y 2022 Precise crop classification of hyperspectral images using multi-branch feature fusion and dilation-based MLP *Remote Sens.* **14** 2713

[17] Zhao J, Wang L, Yang H, Wu P, Wang B, Pan C and Wu Y 2022 A land cover classification method for high-resolution remote sensing images based on NDVI deep learning fusion network *Remote Sens.* **14** 5455

[18] Bouguettaya A, Zarzour H, Kechida A and Taberkit A M 2022 Deep learning techniques to classify agricultural crops through UAV imagery: a review *Neural Comput. Appl.* **34** 9511–36

[19] Sykas D, Sdraka M, Zografakis D and Papoutsis I 2022 A sentinel-2 multiyear, multicountry benchmark dataset for crop classification and segmentation with deep learning *IEEE J. Sel. Top. Appl. Earth Obs. Remote Sens.* **15** 3323–39

[20] Siesto G, Fernández-Sellers M and Lozano-Tello A 2021 Crop classification of satellite imagery using synthetic multitemporal and multispectral images in convolutional neural networks *Remote Sens.* **13** 3378

[21] Li Q, Tian J and Tian Q 2023 Deep learning application for crop classification via multi-temporal remote sensing images *Agriculture* **13** 906

[22] Rustowicz R 2017 Crop classification with multi-temporal satellite imagery *Int. J. of Eng. Res. and V9(06)* **9** 221–5

[23] Zhong L, Hu L and Zhou H 2019 Deep learning based multi-temporal crop classification *Remote Sens. Environ.* **221** 430–43

[24] https://hatarilabs.com/ih-en/sentinel2-images-explotarion-and-processing-with-python-and-rasterio

[25] Jiang G and Zheng Q 2022 Remote sensing recognition and classification of forest vegetation based on image feature depth learning *Mob. Inf. Syst.* **2022** 9548552

[26] Paul S and Nagesh Kumar D 2021 Transformation of multispectral data to quasi-hyperspectral data using convolutional neural network regression *IEEE Trans. Geosci. Remote Sens.* **59** 3352–68

[27] Moumni A and Lahrouni A 2021 Machine learning-based classification for crop-type mapping using the fusion of high-resolution satellite imagery in a semiarid area *Scientifica (Cairo)* **2021** 8810279

[28] https://github.com/bhavesh907/Crop-Classification

[29] Khan A, Vibhute A D, Patil C H and Mali S 2023 Spectral unmixing for end member extraction and abundance estimation *2023 Int. Conf. on Intelligent and Innovative Technologies in Computing, Electrical and Electronics (IITCEE)* (Piscataway, NJ: IEEE) 653–8

[30] Paoletti M E, Haut J M, Plaza J and Plaza A 2019 Deep learning classifiers for hyperspectral imaging: a review *ISPRS J. Photogramm. Remote Sens.* **158** 279–317

[31] Ahmad M, Shabbir S, Roy S K, Hong D, Wu X, Yao J, Khan A M, Mazzara M, Distefano S and Chanussot J 2022 Hyperspectral image classification — traditional to deep models: a survey for future prospects *IEEE J. Sel. Top. Appl. Earth Obs. Remote Sens.* **15** 968–99

[32] https://sentinel-hub.com/explore/eobrowser/

[33] https://indatalabs.com/blog/ml-in-agriculture

www.ingramcontent.com/pod-product-compliance
Lightning Source LLC
Chambersburg PA
CBHW071957220326
41599CB00032BA/6101